F. Walker

Brickwork

A practical Treatise. Third Edition

F. Walker

Brickwork
A practical Treatise. Third Edition

ISBN/EAN: 9783337106614

Printed in Europe, USA, Canada, Australia, Japan

Cover: Foto ©ninafisch / pixelio.de

More available books at **www.hansebooks.com**

BRICKWORK:

A PRACTICAL TREATISE

EMBODYING THE GENERAL AND HIGHER PRINCIPLES OF

BRICKLAYING, CUTTING, AND SETTING

WITH

THE APPLICATION OF GEOMETRY TO ROOF TILING,
REMARKS ON THE DIFFERENT KINDS OF POINTING,
A DESCRIPTION OF THE MATERIALS USED BY THE BRICKLAYER

AND

A SERIES OF PROBLEMS IN APPLIED GEOMETRY

By F. WALKER

CERTIFICATED BY THE SCIENCE AND ART DEPARTMENT IN BUILDING CONSTRUCTION,
PRACTICAL, PLANE AND SOLID GEOMETRY, ETC.

ILLUSTRATED WITH NINETY-ONE WOODCUTS

Third Edition

LONDON
CROSBY LOCKWOOD AND SON
7, STATIONERS' HALL COURT, LUDGATE HILL
1891

[*All rights reserved.*]

LONDON:
PRINTED BY J. S. VIRTUE AND CO., LIMITED,
CITY ROAD.

PREFACE.

THE object of this little work is to give the young artisan a general and practical insight into his trade, and to inspire him with a wish to become a useful and successful workman; which means that he must work with his head as well as with his hands. The greater portion of the matter contained herein is such as to be indispensable to the proficient workman. Though the work does not profess to be in any way an exhaustive treatise on a trade so varied as that of the brick-layer, yet the writer hopes that it may be a help to those who, through the division of labour or otherwise, have had their practice confined to one branch only of their trade; and that it may not be considered altogether unworthy the notice of professional men, being to some extent the outcome of twenty-two years of practical experience in building operations. It is, however, intended chiefly for that large majority of young men who enter the trade of the bricklayer (and all other trades in house-building) without any previous training or instruction to fit them for the calling, depending entirely upon the manipulative skill they may or may *not* acquire in the handling of their tools. The book commences with the site of a building, and goes through the successive stages of the bricklayer's trade, including roof tiling; and concludes with a section on Applied Geometry, containing problems that may be useful in every-day practice.

LONDON, *September*, 1884.

NOTE TO THE SECOND EDITION.

THE very rapid and gratifying sale of the first edition, and the favourable manner in which it has been received by the various technical journals, have led the author to make several additions and a few alterations to the work, with a view to increasing its usefulness not only to the operative student, but also to those who may be preparing for the Science Examination in Building Construction.

CONTENTS.

SECTION I.

MATERIALS AND GENERAL PRINCIPLES OF CONSTRUCTION.

	PAGE
Site	1
Establishing a Level or Datum	2
Setting out Building	2
Concrete	5
Cement	10
Drains	11
Mortar	14
Red Brickwork	14
Bricks	16
Characteristics of Good Bricks	19
Bond of Brickwork	20
Old English Bond	21
Bond of Footings and Walls	22
Setting out the Bond	26
Heading Bond	28
Templates and Strings	30
Bats	30
Flemish Bond	31
Various Bonds	34
Herring-bone Bond	36
Dutch Bond	37

	PAGE
Keeping the Perpends	39
Toothings	39
Grouting	40
Flues	41

SECTION II.
ARCHES IN GENERAL.

Arches	46
Relieving Arches	48
Plain Arches	49
The Skew or Oblique Arch	49
Skew Arch at Brondesbury	52
Water Conduit	56
Groined Vaulting	58

SECTION III.
GAUGED-WORK AND ARCH-CUTTING.

Gauged Work	61
Setting	63
Drawing and Cutting Arches	64
The Bulls-eye	65
Semi and Segmental Arches	66
The Camber Arch	67
The Gothic Arch	69
The Ellipse Gothic Arch	72
The Semi-Ellipse Arch	72
The Venetian Arch	74
The Scheme Arch	75
The Semi-Gothic Arch	76
Gothic on Circle Arch	77
To Find the Soffit Mould	78

CONTENTS.   vii

SECTION IV.
ORNAMENTAL BRICKWORK.

	PAGE
The Niche	79
The Niche Mould	83
Moulded Courses	83
Ornamental Arches	84
The Oriel Window	85
Ornamental Gable or Pediment	87
Gothic Window	88

SECTION V.
ROOF-TILING, POINTING, ETC.

Tiling	92
Roofs having different Pitches	94
To obtain the necessary Angle of Hip or Valley Tiles	96
Pointing	97
Flat-Joint Pointing	98
Burning Clay into Ballast	100
Building Additions to Old Work	102
Fire-proof Floors	102

SECTION VI.
APPLIED GEOMETRY.

To draw a square whose superficial area shall equal the sum of two squares whose sides are given	103
To draw a right-angled triangle, base 1½ inches, height ½ inch	104
To draw an arc by cross-sectional lines	105
To describe a flat arc (camber for instance) by mechanical means	106

CONTENTS.

	PAGE
To find the joints of a flat arch without using the centre of the circle of which the arc is a part	106
To draw the joints of a semi-ellipse arch with mathematical accuracy	107
To find the invisible arch contained in a camber	108
Any two straight lines given to determine a curve by which they shall be connected	109
To find the form or curvature of a raking moulding that shall unite correctly with a level one	111
To describe an ellipse by means of a carpenter's square and a piece of notched lath	112
To draw a Gothic of any given height and span; or, in other words, an Ellipse Gothic	113
To draw the arch bricks of a Gothic arch, that is for the curve in the previous problem	114
To find the radius of any arc or arch, the rise and span being given	114

BRICKWORK.

SECTION I.

MATERIALS AND GENERAL PRINCIPLES OF CONSTRUCTION.

SITE.

THOUGH the bricklayer is very seldom called upon to choose the site of a proposed building, he should nevertheless make himself acquainted with the essentials of a good foundation, and the characteristics of a bad one, as a subject not altogether foreign to his calling. The workman who rests satisfied with just the manipulative knowledge of his own trade is not likely to realise the value of the word *progress*, and must of necessity be content to remain in the position in which he found himself placed as a workman. Though the bricklayer has no voice in the choice of site, he may, as foreman or clerk of works, have to a great extent the power of minimising the evil effects of a bad one, if he be possessed of the necessary knowledge. For be it remembered that a good foundation is as necessary to the stability of a building, as good flues and drains are to the health and comfort of its occupants. The best sites to build upon are hard gravel, igneous and metamorphic rocks, limestones, sandstones, and chalk. A clay foundation should be well drained, as clay by its impervious nature retains moisture, and the whole area of the site covered with 6 inches of surface concrete, made up with Portland cement or ground blue lias lime, to keep back ground-damp, which will otherwise be attracted by the warm air within the building. When building on a clay or sand foundation the building should be kept level throughout, as by building up one portion of the building and leaving down another, ugly fractures sometimes occur in the walls, caused by one portion of the work settling at one time, and other portions at another, which greatly mar the appearance of the structure.

Establishing a Level or Datum.

Before excavating trenches to receive concrete for footings, a level, or *datum* as it is technically called, should be established. To do this, drive a large stake well into the ground where it will not be likely to get disturbed, and let the top of it be the ground-floor level, which must be taken off the drawings if not otherwise determined. To avoid the possibility of mistakes, all levels for excavations, concrete, and brickwork should be taken from this only.

Setting out Building.

In setting out a building, one or other of the following methods is generally adopted. Either the extreme side walls are squared from the line of frontage, which is given, and the positions of the intermediate walls established by parallels; or, two centre lines are drawn at right angles, right through the plan of the building, and the walls set out at parallel distances from them; taking all measurements from the centre lines. The positions of walls should not be laid down by measuring the distance of one wall from another in succession; for if an error be made in the setting out of the first wall, it will, in this way, be perpetuated from one wall to another throughout the building. But by measuring from the centre line, an error would be confined to that particular wall in connection with which it was made, and would be readily discovered when checking the distances between the respective walls. In both methods we have assumed the building to be square. If the setting out is to be

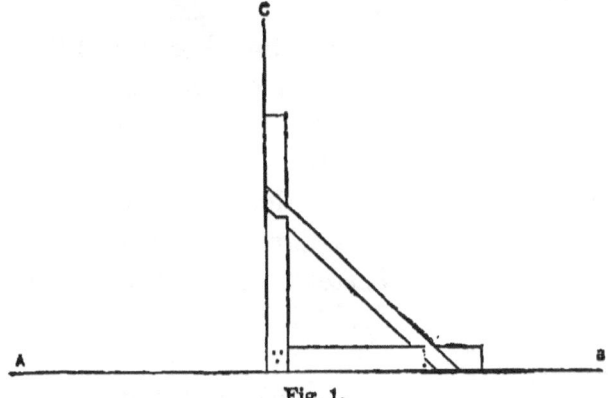

Fig. 1.

done by means of a large square, which is generally the case, it should be tested or proved before use.

To do this, draw a line $a\ b$ along a straight

edge (Fig. 1), not less than twice the length of the base of the square. Adjust the base of the square along this line from *b*, and draw a line *c* along the perpendicular blade until it meets the base line *a b*; now reverse the square along the base line from *a*, and if the square be true its perpendicular will coincide with the perpendicular line *c*. Another way of setting out the side walls from a given line of frontage is by means of a 10-feet rod. Having drawn a line tightly to represent the front of the building, along this line measure 6 feet from the quoin (French *coin*, a corner), and push through the line at the 6-feet point an ordinary brass pin. Draw another line in the same way as the first, approximately at right angles to it, and from the quoin again measure off 8 feet along *this* line, fixing another pin as before at the 8-feet point.

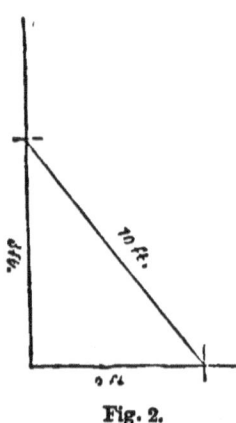

Fig. 2.

With one end fixed at the quoin, the other end of the line must be moved until there be a distance of 10 feet between the two pins measured across the angle. The lines will then be square one with the other. Instead of 6, 8, and 10, we could have taken 12, 16, and 20; but whatever figures be used must stand in the same ratio or proportion to each other as the above, and shown in Fig. 2.

Another Method.—From point B (Fig. 3), with

steel measuring tape set off 30 feet, or more or less as convenient, at an approximate angle of 45 degrees with the given line A B. From D mea-

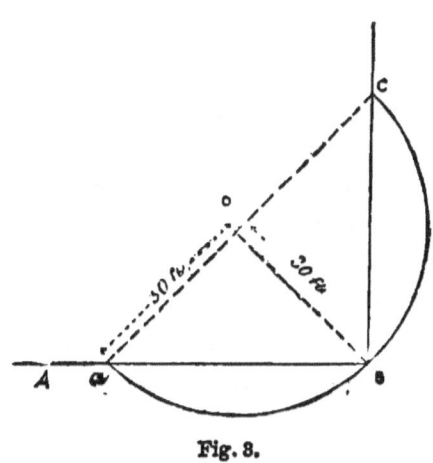

Fig. 3.

sure off the same distance to *a*; from *a* draw a line through D, measuring from D to *c* 30 feet. A line drawn from B through *c* will be at right angles to the given line A B, the line of frontage; B would be the quoin of building. This depends upon the principle that all triangles in a semicircle are right-angled triangles, and all the angles in the same segment of a circle are equal (*Euclid*, bk. iii. prob. 21).

CONCRETE.

The thickness for concrete varies from 1 to 3 feet, according to the nature of the subsoil upon which the building will stand; but in some cases it is very much thicker, as in made-up ground, where, to ensure a good foundation, it is necessary to go down to the London clay, or some other firm substratum, depending upon the nature of the ground. The Metropolitan Building Act requires that the concrete shall not be less than 9 inches in depth, nor have a margin of less than 4 inches outside the first course of footings; 6 inches is the usual margin in good work.

The following is a specification to govern the supply of materials, the mixing, and the putting into place of cement concrete. The whole of the cement to be Portland of the very best quality, very finely ground, weighing not less than 110 lbs. to the striked bushel, of which 90 per cent. must pass through a sieve of 2,500 meshes to the square inch, and it must be capable of maintaining a breaking weight of 350 lbs. per square inch, after being made in a bronze mould immersed in water during an interval of seven days.

The mixing to be carried on upon a clean platform made of 9 inch × 3 inch deals, bedded solidly on sand, that the cement may not run off through the joints in the process of mixing. The concrete to be composed of four parts of broken bricks, broken porous stone, or Thames ballast; two parts sharp clean sand, free from loam or other impurities; and one of cement of the specified quality. The parts to be measured in a half-yard cubic box (3 feet × 2 feet × 2$\frac{1}{4}$ feet), and thoroughly mixed together in a dry state. The ballast or broken bricks to be capable of passing through a 2-inch mesh. The dry concrete to be heaped up and turned over at least twice before wetting. The water to be applied through a rose, not more to be used than is necessary to mix the whole very thoroughly. While the water is being sprinkled on, the mixture should be drawn down by "picks," while two or more other men turn it over, after being so drawn down, to another part of the platform, from which it must be again turned over until the parts are thoroughly incorporated. The concrete to be tipped from a height not exceeding 4 feet, and to be steadily rammed or struck with the back of a shovel until the cement or matrix flushes to the surface. The whole to be left solid and clean.

In the treatment of concrete much depends upon experience and judgment, and it is therefore the more difficult to lay down hard and fast rules to govern the proportion of the ingredients and the mixing of them. The one thing to be aimed at in the apportionment of the ingredients is homogeneity; where this does not exist, strength will be wanting.

As regards "packing," or the practice of placing stones or other suitable material larger than the aggregate, in the mass of the concrete, it is objectionable under certain conditions. In a thoroughly good Portland cement concrete, if properly treated, there will neither be contraction nor expansion to any perceptible degree in the setting; and in such there is no objection to packing, if the stones or other material be uniformly distributed and solidly bedded in the mass. But in an inferior concrete subject to contraction or expansion, packing is decidedly objectionable, and likely to lead to injurious results; more especially if the packing be not evenly distributed throughout the concrete. This consideration has led engineers and architects to adopt in their specifications the precautionary clause that the aggregate shall be of an uniform size—generally, to pass through a 2-inch or $2\frac{1}{2}$-inch ring.

The quantity of water to be used depends almost entirely upon the nature of the aggregate; ballast or any siliceous aggregate requiring only enough to thoroughly mix the cement, while that of a porous nature, such as broken bricks, would require more. The proportion of cement must be governed by circumstances, for while the Metropolitan Main Drainage Works adopted one of cement to five and a-half of aggregate, we are informed by Mr. Reid *On Concrete*, that in the sea forts of Copenhagen the concrete was made in the following proportions:—

Portland cement	1
Sand	4
Fragments of stone	16

and the concrete for filling in the terra-cotta at St. Paul's School, Kensington, consisted of one of Portland cement and ten of aggregate.

In Portland cement concrete, "a rotten or friable material is to be avoided, except where unavoidable, and in that case only in combination with a large quantity of cement, so as to neutralise as far as possible any tendency to weakness. Sand, where a choice exists, should be as rough and coarse as possible, and that made by the various natural or physical influences from sandstone, limestone, or other similar rocky formations, is to be preferred over those from flint or volcanic rocks. The former sands or shingles are more porous than the latter, and consequently better able to absorb the silicates of the cement when being mixed. For this reason it is advisable not to have the sand, gravel, or shingle too fully saturated with water; if this is so, the matrix is unable to imbibe the fluid portion of the mixture, and consequently it is thrown off as waste from the concrete. This observation equally applies to the mischievous practice of over-wetting bricks in building with cement mortar. A dry brick is bad enough, but when saturation is carried to excess equally faulty results ensue. With regard to the acting properties of Portland cement when used with salt sand, or salt water, an experiment proved the use of salt water and salt sand perfectly satisfactory, both with Portland cement and lias lime, but there was no question as to their setting being retarded by their use."—*Brunel.*

When blue lias is used for concrete, the proportion of parts and the mixing is the same as described in cement concrete.

Burnt ballast is frequently used as an aggregate for concrete, but care should be taken that it be thoroughly burnt free from clay. Burnt ballast concrete should be made rather sloppy on account of its absorbent nature, or it will quickly absorb the moisture from the cement or lime with which it is mixed, to the injury of its setting properties and ultimate strength.

Mixed with one-third of Thames ballast and a fair proportion of lime it will yield a good concrete for footings to walls.

Cement.

Adie's No. 1 cement testing machine is very generally used for testing cements, but where one of these is not at hand they may be roughly tested in the following manner. Having mounted a briquette (Fig. 4), whose sectional area is one square inch, or more as the case may be, after seven days' immersion let it be suspended from one end, and from the other end suspend a cement barrel containing sand, increasing the quantity until the briquette breaks or its power of resistance be overcome. The sand should not be thrown into the barrel, but slid into it by means of an inclined plane, and in small quantities. The weight of the cask with its contents will represent the breaking weight. With Adie's machine the briquette in the making is subjected to a slight pressure, which adds considerably to its tensile strength, so that the resistance to breaking of a briquette made by the machine will be greater than that of a briquette of the same cement made by hand and not subjected to pressure. Another way : bed two bricks together (Fig. 5), and after a few days' immersion let them be suspended and treated in the same way as the briquette. This plan is suitable for ascertaining the comparative strength of cements, but in so doing the same kind of bricks, sand (if any used), and even water should be used, and the exact proportions maintained in the mixing, or, in other words, the conditions should be exactly the same. Bricks having a smooth impervious bed will be found to have less adhesion than those of a hard but comparatively porous nature—pressed bricks and hard stocks, for instance.

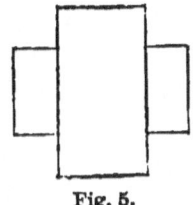

Fig. 5.

The bricklayer should make himself acquainted with the various limes and cements, and the ingredients used in combination with them; also with concrete, as subjects belonging particularly to his trade, and which by reason of his occupation he has a better opportunity of doing than any other class of operatives. In large and important public works these are generally subject to the inspection of a bricklayer.

Drains.

The laying of drains, at once the most important and too frequently the most neglected part of a building, should never be intrusted to unskilled workmen. The fall having been determined, which should not be less than one in sixty or one inch in five feet, the flange of each pipe should rest upon a bedded brick, that the joints may be caulked all round with gaskin or oakum previously to being made up with Portland cement. The object of caulking is to prevent the cement squeezing through into the pipe, a very common cause of stoppage in drains. They can now be bedded half way up in fine concrete, so as to form a cradle, care being taken not to disturb the joints. The inside joint of each length of pipe as it is laid should be stopped with Portland cement, and left solid and clean, free from anything approaching to burrs. The drains should be laid down air and water-tight, free from "dips," with no right-angled junctions nor sharp bends, and kept, if at all possible, outside the building, with inspection holes large enough for a man to work forcing-rods in case of a stoppage. A length of pipe in the man-hole should have a movable top. This kind of pipe is called an *operculum* or "channel" pipe. In many instances only the invert half of the pipe is used in that portion of the drain passing through the man-hole, which is ventilated by a current of fresh air entering the man-hole, passing through the entire length of the drains, and finding an outlet through the open soil-pipe above the roof. In such an arrangement a trap should intervene between the sewer and the man-hole, to prevent the possibility of sewer gas escaping through the fresh air inlet. But where fresh air is not introduced, the trap may be dispensed with, the soil-pipe serving as a ventilator both for the sewer and the drains.

Six-inch pipes will be found large enough for most buildings. As the subject of trapping, disconnecting, and ventilating drains belongs to sanitary science, it cannot be further noticed here beyond giving a plan and section of a dip-trap (Figs. 6 and 7) which the bricklayer is sometimes

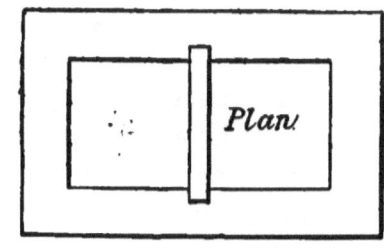

Fig. 6.

Fig. 7.

called upon to build. This trap should be used only where there is a copious and frequent supply of water (but not in connection with soil), as by its size and construction a greater quantity of water is required to trap it than the earthenware traps now more generally and preferably used.

Mortar.

Mortar used by the bricklayer is made either from stone lime, lias, or Portland cement, mixed with a proper proportion of sand. Chalk lime should not be used, as the only setting that takes place in it is the formation of a surface crust, bearing a small proportion to the bulk. Stone, or gray chalk lime, as it is sometimes called, is generally used; it possesses slight hydraulic power, and will set if secluded from the air or in damp situations, and is capable of bearing three parts of sand to one of lime. For damp situations blue lias will be found to make the best lime-mortar. It is eminently hydraulic, and becomes very hard, especially in damp places; but it will not bear so much sand as stone lime. The amount of sand should not exceed twice that of lime. Lump lias is used for mortar; it should be well wetted, covered over with sand, and allowed a day to slack before being ground in the mortar mill. The sand used for all mortars should be a clean, sharp, angular grit. Cement has been already spoken of in connection with concrete, and elsewhere.

Red Brickwork.

Owing to the revival of the Queen Anne style of architecture, brickwork now occupies the foremost position in building construction, of which very good samples may be seen at Westwood House, Sydenham; Fitz-John Avenue, Hampstead; the Chelsea Embankment, and many other places in and about London. Our popular architects delight to revel and indulge their fancies in red brickwork, as evidenced in several public buildings of recent erection. The Victorian age, from an architectural point of view, will be conspicuous for its stuccoed buildings and its red brickwork—the former an expressionless imitation, the offspring of the speculator, and the Caliban of architecture. But Truth in architecture, as in all things, *will* assert herself; she breathes into the nostrils of a second Adam, and lo! we have "a thing of beauty."

We can remember, in our experience, when the life of the bricklayer was often made "bitter with hard bondage in mortar and in brick," by reason of the reign of stucco; but, thanks to the able advocacy of Mr. Ruskin and the late Mr. E. Street, such rapid strides have been made in brickwork that one is almost surprised to see the amount of art-workmanship wrought in red-brick designs.

These will be found mostly in retired out-of-the-way streets, relieving, both by colour and detail, the dull monotony of the unbroken line of our vista-like old street architecture.

Some years ago the Philological School, St. Marylebone Road, was pointed out as a sample of ornamental brickwork. The ornamental features in this structure are made up of a judicious use and arrangement of polychrome bricks, and stone dressings. The building is, undoubtedly, a good one, possessing that repose almost peculiar to ecclesiastical architecture. But the term ornamental brickwork is so closely associated in these days with the idea of *form*, that we are accustomed to exclude from the meaning of that term all brick designs characterized by an absence of projection.

We know no better samples of red brickwork than St. Paul's Schools, and the City Guilds Technical Institute, Kensington; and the Midland Hotel, St. Pancras Station.

Bricks.

In dealing with brickwork it is necessary that something should be said about bricks, though it is not intended to go into the chemical properties or other scientific matters connected with them, as we are presumably writing for persons in or connected with the trade of a bricklayer, but will just take a passing glance at the bricks commonly used in and about London, and state the purposes for which they are best adapted.

Stock bricks are divided into "picked" stocks (picked for colour and hardness), "washed" stocks, "grizzles," "place," and "shuffs." "Shuffs" are worthless, "place" are little better; "grizzles" are those bricks which have a good face or end with the other face or end underburnt, and similar in appearance to "place," which are of a reddish colour. "Picked" are those which are suitable for good exterior facing. "Washed" stocks, on account of their softness, are fit only for interior facing. The best stock bricks for general facing purposes are those called "shippers," which, as their name implies, are sorted for shipping.

Malms are a superior kind of stock bricks, made of washed clay and chalk, and are used for superior facing and for "cutting" purposes, but are not suitable for "gauged-work" on account of the numerous small air-cells contained in the bricks, which make it impossible to rub them up to an arris, which is indispensable to good setting.

Of red building bricks there are a great variety in the London market, the best of which for colour and weathering properties are Fareham reds, though rather irregular in shape. St. Thomas's Hospital, and the Nurses' Training Home, Queen Anne's Gate, St. James's Park, are faced with these. Sometimes they are rubbed down to obtain true faces; but this should be avoided for the sake of preserving the deep red colour, which constitutes the beauty of these bricks. Fareham rubbers for "gauged-work" also stand first in quality, though they are not extensively used, as they are dearer than the other varieties in the market.

Next in quality come the Berkshire Builders and T. L. B. Rubbers, made by T. Lawrance, Bracknell, Berks. The Teynham bricks, stamped G. Richardson, Teynham, are good bricks, possessing in a large degree the qualities that recommend the Farehams, and with the additional advantage of a fairly good shape. Gault bricks are much used for facing; they are much harder than stocks, and also dearer. Of white bricks Suffolks are the very best. They are a close, firm brick, suitable for first-class facing, either exterior or interior, or for "gauged-work." They are of a soft nature, but harden very much by exposure to the action of the atmosphere.

A very nice piece of work—three-light geometrical windows—executed in these bricks, and designed by Messrs. H. Saxon Snell and Sons, 22, Southampton Buildings, W.C., may be seen in the chapel attached to the Rackham Street Infirmary, Notting Hill, W. Staffordshire blue bricks are the most suitable for external bases, plinths, and dwarf-walls for palisading, or wherever there is much traffic.

Enamelled bricks are now very extensively used instead of tiles; they can be obtained in various colours, and are suitable for facing dairies, &c., and areas where reflected or borrowed light is required. They are obtainable in double headers,

Fig. 8.

viz. two ends enamelled for 9-inch walls, and double stretchers for 4½-inch walls, single headers and stretchers for facing, and bullnose and chamfered bricks (Fig. 8) for jambs or reveals. The best kind are those bearing the stamp, "Cliff, Wortley, Leeds."

Firebricks should be used for all places exposed to the action of fire or intense heat. They are made of fireclay, and should be set with close joints in a mortar made of the same material, wetting the bricks before setting them. The mortar under the action of the fire will vitrify, and form one body with the bricks. In lining boiler furnaces, &c., bricklayers frequently use fireclay only with that portion of the work that will be subjected to the flame, but it may be set down as a rule that wherever it is necessary to use firebricks, it is also necessary to use *fireclay* to bed them in. Nevertheless, when it is not readily obtainable, plaster of Paris and sand may be used as a very good substitute for small jobs, but on no account should cement be used, for being non-elastic it will fracture under the action of intense heat. Stourbridge bricks are much used as the best kind of ordinary fire-bricks, but Dr. Siemens has shown the Dinas firebricks to be the best, and to be capable of resisting the temperature of 4,000° to 5,000° Fahr.*

CHARACTERISTICS OF GOOD BRICKS.

Soundness, freedom from flaws, cracks, or stones of any kind. They should contain no lumps of lime or limestone, however small; should be regular in shape and uniform in size, their length exceeding twice their breadth by the thickness of a mortar joint. They should not absorb at most more water than is equal to one-sixth of their dry weight. They should be hard, and burnt so thoroughly that there is incipient vitrification all through the brick. When struck together they should yield a clear metallic ring. (This last-

* Dr. Siemens' "Chemical Society," 7th May, 1868.

mentioned characteristic belongs more to stocks and the harder kind of bricks.) Their texture should be homogeneous and compact. They should be regular in colour, with their arrises square, sharp, and well-defined. Pressed bricks, such as those from the midland counties and Ruabon, are almost non-absorbent, and for all practical purposes impervious to water. The nearer bricks approach to imperviousness the better will they be.

The following is an analysis of the clay worked by Messrs. Monk, Newell, and Bryon—Ruabon—

Moisture	1·54
Combined water	3·54
Silica	63·00
Alumina	18·0
Sesquioxide of iron	6·70
Protoxide of iron	1·95
Potash	2·37
Soda	3·10
	100·20

Bricks and terra-cotta, manufactured from this clay, may be seen at the Northern Hospital, Winchmore Hill, London, now in course of erection by Messrs. Wall Brothers, of London.

Bond of Brickwork.

We will now enter into what might be termed the scientific part of bricklaying, and it will not be out of place to repeat what Smeaton wrote half a century ago with reference to this subject, and which is equally true to-day: "As the art of bricklaying is generally supposed to be so simple as to require little or no attention, it will be necessary to remove this false impression by a somewhat particular detail of the facts which relate to it. There are many persons, and even some workmen, who suppose that nothing more is required than that the bricks should be properly bedded and the work level and perpendicular. But the workman who would attain perfection in his business should acquaint himself with the different arrangements made use of in placing [bonding] the bricks, so that one part of the work shall strengthen another, and thus prevent one portion from a greater liability to give way than another."

So much for the statement of an eminent engineer, than whom none knew better the value of bonding, as evidenced in the old Eddystone Lighthouse, which was so thoroughly bonded, one stone into another, and each into the whole, that nothing but the wearing away of the rock upon which it stood led (or was likely to lead) to its demolition.

Old English Bond.

Old English bond consists of alternate courses of headers and stretchers, while Flemish bond consists of alternate headers and stretchers in each course. Old English is the only *true* bond, the other bonds (and there are several) being merely arrangements to please the eye. Gwilt, referring to bond, remarks, in his "Encyclopedia of Architecture," that "previous to the reign of William and Mary all the brick buildings in the island were constructed in what is called English bond; and subsequent to the reign in question, when in buildings as in many other cases Dutch fashions were introduced, we regret to say much to the injury of our houses' strength, the workmen have become so infatuated with what is called Flemish bond that it is difficult to drive them out of it. To the introduction of the latter has been attributed (in many cases with justice) the splitting of walls into two thicknesses; to prevent which expedients have been adopted which would be altogether unnecessary if a return to the general use of English bond could be established."

Bond of Footings and Walls.

The Metropolitan Building Act requires that the footings of all walls shall not be less than twice the thickness of the superincumbent wall, or, as bricklayers call it, "the neat work."

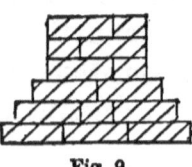

Fig. 9.

Fig. 9 represents the footing for a brick-and-a-half wall. A two-brick wall would require a four-brick footing, and so on, according to the size of the wall, setting back $2\frac{1}{4}$ inches on each course of footings until the wall be brought into its proper size. Where a "bat" occurs in the footings, as in the second course, it should always be kept in the centre. Fig. 10 shows in elevation the footings and three courses of a 14-inch wall. It will be seen that the "closer" is not used until the

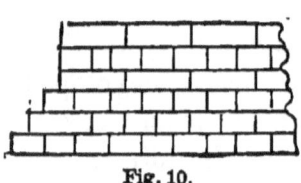

Fig. 10.

setting out of the bond for the "neat work." Figs. 11 and 12 are the plans of two successive courses of a one-and-a-half brick wall, showing the sectional bond. It will be seen by this that there are no two joints in the wall immediately one above the other, but that in the direction of the length of the wall there is a lap or bond of $2\frac{1}{4}$ inches of each brick over the two immediately below it in the next course, and a lap of $4\frac{1}{2}$ inches in the width of the wall. This result is obtained by running the transverse joints right through the

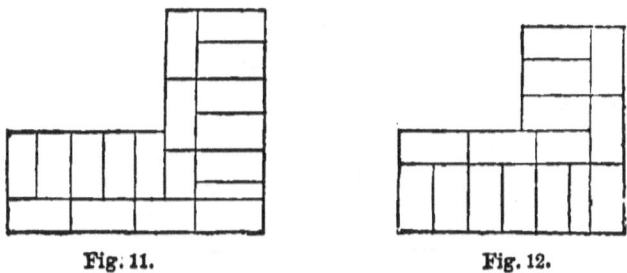

Fig. 11.   Fig. 12.

wall from one side to the other. A simple principle, but seldom carried out even by bricklayers.

The method in general practice is shown in Figs. 13 and 14. It will be seen that the transverse or "cross" joints do not run through the wall, but that the ends of the stretchers come in the middle of the headers, consequently the cross joints in the middle $4\frac{1}{2}$ inches of the wall are one over the other from the bottom to the top of the wall. This is caused by showing full "stretchers," *a* and *b*, in the internal angle, instead of letting them pass $2\frac{1}{4}$ inches into the return wall, as in Figs. 11 and 12. Many bricklayers insist upon showing a whole *"stretcher"* in the angle in all cases; but he who insists upon this has

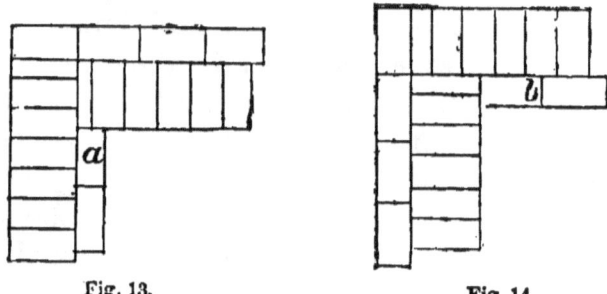

Fig. 13.   Fig. 14.

yet to learn the bond of brickwork. The reader would be greatly helped to an understanding of bond by having a few model bricks, and arranging

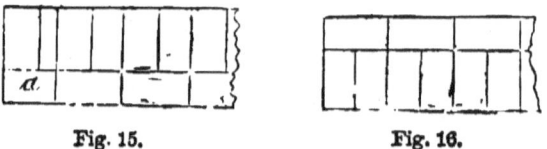

Fig. 15.   Fig. 16.

them as shown in these figures. Figs. 15 and 16 represent a straight jamb in a 14-inch wall. Here again, that the "cross" joints may run straight through the wall, it is necessary to introduce a

Fig. 17.   Fig. 18.

three-quarter "stretcher" *a*, and to omit the "closer" in the next course above. Figs. 17 and 18 are the plans of two consecutive courses of a pier 14 inches on the face and 18 inches deep. The face bond is made up of two three-quarter "stretchers" on one course, and of three "headers" on the other. Figs. 19 and 20 are two courses of

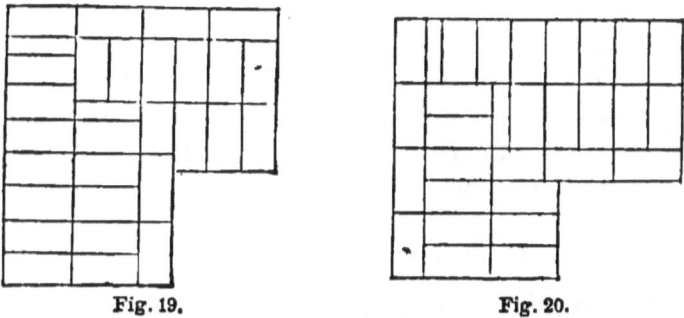

Fig. 19.   Fig. 20.

a wall two and a half bricks thick. In all walls of such a size as to take an odd half brick (two bricks and a half, three bricks and a half, &c.), the "stretcher" is always laid on the outside face in one course and on the inside face in the next course.

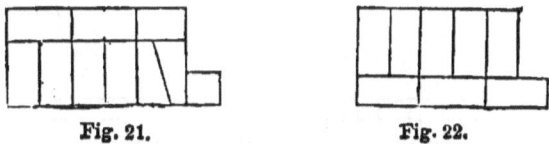

Fig. 21.   Fig. 22.

Figs. 21 and 22 show the "king closer," which in practice, owing to the trouble of cutting and the probability of breaking in the cutting, is seldom used. In this case two bricks are cut in their whole length from $2\frac{1}{4}$ inches to $4\frac{1}{2}$ inches, but it is more frequently cut out of one brick, as in Fig. 23, and an adjoining "bat" is cut to fit it.

Fig. 23.

A great many instances of bond in different

sized walls and piers might be given, but as a thorough knowledge of "bonding" can be obtained only by practice, we will not multiply examples.

If the bricklayer adhere to the principle of keeping the "cross" joints immediately opposite each other, and laying the bricks in one course quarter bond with the bricks in the course below it, he will experience little difficulty with any sized wall or pier.

Setting Out the Bond.

The chief thing in connection with brickwork is setting out the bond, for which a good bricklayer should be selected. This will be more readily conceded when we consider the strains to which a building is subject. The bond should be

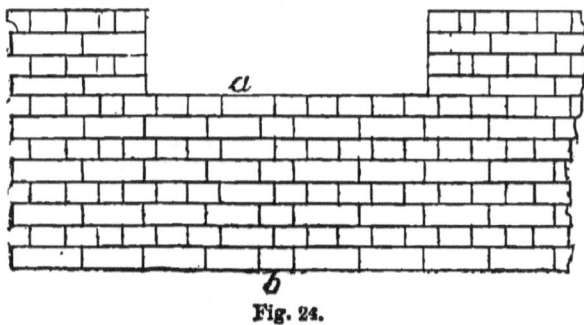

Fig. 24.

set out at least one course below the ground line, and the positions of doors, windows, panels, or large apertures taken off the drawings. This is best done in a stretching course, setting a "perpend" for every reveal or jamb, and working the "broken bond" under each window, or other aperture, as the case may be, as in Fig 24, *a* and *b*. Reveals and jambs in point of bond should be treated as "quoins." Where a base occurs the "bond" should be so arranged that a whole brick will work in the internal angle above the plinth.

In Fig. 25 (plan and elevation) we have a 2¼-inch plinth; a "perpend" or vertical joint in the stretching course is started 6¾ inches from the angle at the base; this joint "plumbed" up will be 9 inches, or a brick, from the angle above the plinth, and work proper or conventional "bond." In many cases the base is treated by bricklayers as if it were a detached part of the building,

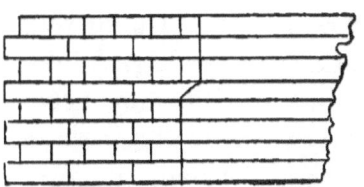
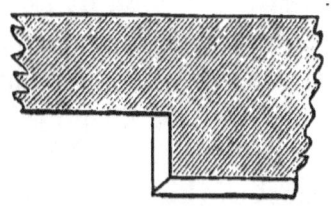

Fig. 25.

and the consequence is that "closers" are to be seen in the internal angles of many good buildings where whole bricks should be found. Such things, though small in themselves, go a long way to make up or to detract from the general effect and appearance of brickwork.

"Broken Bond" is the result of badly proportioned piers; thus, in a pier 3 feet 2½ inches long, the bricklayer would have to work four bricks and a quarter, but to do away with the quarter or "closer," a header and a three-quarter "stretcher" are substituted for a "stretcher" and the "closer," the three-quarter and "header" making up the "broken bond," and are kept as near as possible in the middle of the pier.

The work once above the ground, the building should be levelled all round, and a piece of hoop-iron fixed in a joint at each corner or angle to gauge or measure from, taking care that they are all in the same level course. A "gauge-rod," reaching from floor to floor, with all the courses and stone strings (if there be any) and heights of window sills and heads marked on it, should be given to the bricklayer to work to, by which means he can at any time see how his work is rising, which in London should not exceed nor be less than four courses to a foot; and the careless or inferior workman will then have no excuse for not keeping his work level and to the gauge. Not working to a gauge-rod is the chief cause of thick and thin joints, though any competent workman with a 2-feet rule should be able to keep his work right. The bricks in building should be wetted, but not to saturation, and the mortar of such a consistency that the "cross" joints between the bricks can be drawn up as the bricks are laid; any open or partially filled joints can then be filled by "flushing," which is to be preferred to "grouting," and should be done on every course.

Heading Bond

is the name given to that arrangement in which the bricks are laid all "headers." This bond is used in circular and curved walls of a short radius, and in round chimney stacks, so as to keep the wall within the "sweep," or arc, for if "stretchers" be used, every 9 inches of the wall will be a straight line, and when built will consist of projections and hollows, and will be in that state described by bricklayers as *"hatching and grinning."* Heading bond should never be used on straight walls or where it can be avoided, as very little longitudinal strength is obtained, as will be seen by refer-

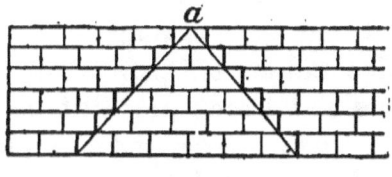

Fig. 26.

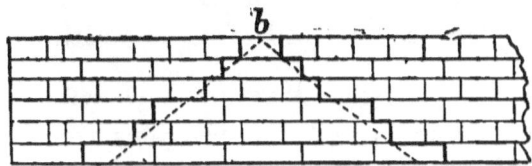

Fig. 27.

ence to Figs. 26 and 27, showing the angles of strain in two walls, one in heading bond and the other in English. The thick lines show the direction a fracture would take in the event of a settlement. They also show the space over which any given weight resting on either *a* or *b* would be distributed; and this idea leads us to the consideration of the use of stone templates and strings in connection with brickwork.

Templates and Strings

Templates under girders, principals, beams, &c., should always be of York, never of Portland or any similar stone, and should be at least 14 inches long—18 inches would be better, but the length must be regulated by the weight which it has to carry. There is little doubt that " string courses " in the shape of a flush band were first introduced to impart strength to walls whose component parts were of diminutive dimensions (the Roman tile for instance, used in Roman walling), and that their ornamental feature was a secondary idea and an outgrowth of the former. String courses and bands are still used very extensively for this purpose, and are placed generally at the floor line, the window sill level, or the window head or springing line, and in some buildings in each and all of these positions.

Bats.

A consideration of the previous remarks will have illustrated the evil attending the use of " bats." The greatest evil in connection with them is that workmen when walling, instead of fairly distributing them amongst the whole bricks, generally allow them to accumulate on the scaffold, and when they have a quantity put them in the wall all together, much to its injury. Good work may be done with a fair proportion of " bats " if they be used with discretion; and it is only fair to the builder that he be allowed to use the bats made on the job.

FLEMISH BOND.

Having already pronounced upon the merits of this bond and given the opinion of an eminent authority (Gwilt), little remains to be said on this subject beyond explaining a few examples in different sized walls and piers.

Figs. 28 and 29 show a 14-inch wall with a

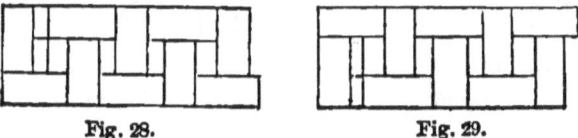

Fig. 28.   Fig. 29.

straight jamb, both sides Flemish bond, showing the way such a wall is generally bonded in practice. The rule laid down to keep the "cross" joints straight through the wall is departed from in this example, consequently the joints in the middle of the wall are one over the other in the entire height of the wall. The proper method is shown in Figs. 30 and 31, in which the "closer"

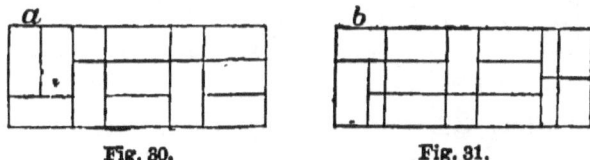

Fig. 30.   Fig. 31.

is dispensed with, and two "headers," *a*, in one course and a three-quarter "stretcher," *b*, in the other are used. A heading and stretching course are obtained by laying whole headers on one face and "snapped headers" on the other. A still better bond would be obtained by laying the headers on each face, alternately "header" and "snap;" but to prevent all "snaps" coming over each other and all whole headers over each other they (the "snap headers" and the whole "headers") should be alternated in the height as well as on the level.

Figs. 32 and 33, the same wall, with the face

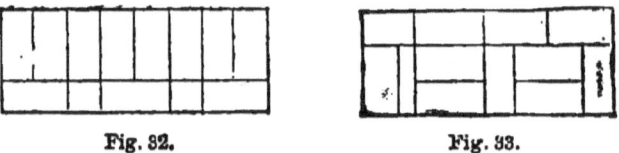

Fig. 32. Fig. 33.

in Flemish and the back in English bond. A good strong wall can be obtained in this way, and where the inside has to be plastered it should always be so built.

Figs. 34 and 35, a two-brick wall, Flemish bond both sides. By snapping the headers in

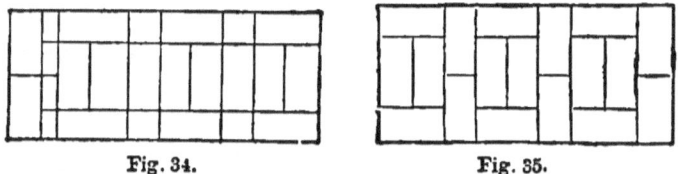

Fig. 34. Fig. 35.

one course, 34, and putting them whole in the other, 35, a heading and a stretching course are obtained, which gives a much better bond through the wall than if all whole headers were used. Fig. 36, a quoin in isometric projection, showing the internal and external angle, and a perfect bond as far as obtainable in Flemish bonding with the inside face built in Old English bond. Fig. 37 gives the bond of a two-and-a-half brick pier projecting from a wall. At *a* is shown a broken bond—two "stretchers" in one course and three "headers"

Fig. 36.

in the next course above them, which frequently occurs, and is the only legitimate "broken" bond in Flemish. Where a three-quarter "stretcher"

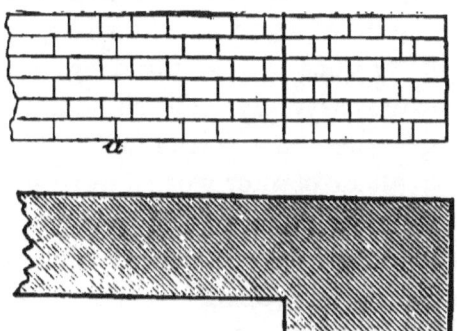

Fig. 37.

occurs as "broken" bond, it can be obviated or done away with by "reversing" the bond on one end of the pier or wall. Thus for a "stretcher" substitute a "header" and "closer."

Various Bonds.

Chimney bond is a term applied only to $4\frac{1}{2}$-inch external walls to chimney stacks. In this arrangement the disposition of the bricks is such as to obtain the greatest possible strength by bonding in the "withes" on every second course, and avoiding the use of bats as far as practicable.

Stacks of $4\frac{1}{2}$-inch walls should never be built in Old English bond, for the reason that bricklayers, when cutting the half bricks to form "snap headers," will sometimes cut them $3\frac{1}{2}$ inches in depth instead of $4\frac{1}{2}$ inches, depending upon the pargetting or mortar to make up the thickness of the wall, which when the flue comes into use will shrink and crack, and falling away from the brickwork leave a stack, in many cases, built partly of closers. English bond is also objectionable on account of the numerous bats. Another practice in $4\frac{1}{2}$-inch stack building, and which cannot be too severely condemned, is that of "buttering" the cross joints with the point of the trowel; or, in plainer words, putting a mortar joint between the ends of the bricks, extending in about 1 inch from the face, the remaining $3\frac{1}{2}$ inches being left open, excepting what little may be filled up in the process of pargetting.

We believe this practice, together with that of plugging into $4\frac{1}{2}$-inch chimney walls for fixing skirtings, to be a fruitful source of many fires, with accounts of which we are occasionally startled. The mortar or cement joints should be put *right through* the width of the bricks, and drawn up solid and tight. Stacks with 4½-inch walls may often be built with advantage in Flemish bond; but the main thing to be attained is strength, which is to be obtained only by bonding in the "withes" or divisions between the flues. Another reason the author would advance in objection to 4½-inch walls for chimney stacks is that plumbers, in "flashing" round the base, cut out the joints for the purpose of turning in the lead; and when wedging the same, thoughtless of the power exerted by the wedge, often break the bond of adhesion between the mortar and the course above the "flashing," leaving the stack in this condition to withstand a wind pressure of from 40 to 50 lbs. on the square foot during a hurricane, often resulting in a coroner's inquest. Zinc "soakers" may be used with much advantage in connection with stacks built with 4½ inch walls, and the angles formed by the junction of the stack and the slating filled in with a small cement fillet, triangular in section, making a perfectly sound and water-tight job, doing away with the necessity of flashings, and preventing the evils that sometimes attend them.

English garden-wall bond consists of three courses of "stretchers" to one course of "headers." This bond may be said to have grown into disuse, excepting in the north of England, where five courses of "stretchers" to one course of "headers" are frequently used in general building. Flemish garden-wall bond consists of three "stretchers" to one "header" in every course, as in Fig. 38. Garden-wall bond is used only, as its name implies, for 9-inch garden walls that have to be kept fair

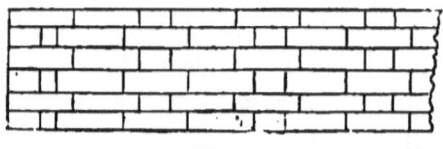

Fig. 38.

or smooth on both sides. The bricks vary most in their lengths; the more "headers" that are put through the wall will, therefore, add to the difficulty of keeping it straight.

Herring-bone Bond.

Figs. 39 and 40 represent a panel filled in with bricks laid "herring-bone." The former is gene-

Fig. 39.   Fig. 40.

rally the method used in paving, where the bricks are laid on their beds, 4½ by 9 inches, in sand, and "grouted" up with cement or mortar. The latter is used for filling in panels under windows and for tympana of arches, and are laid four courses to the foot. When a large area of paving has to be done in this way, the simplest way will be to work from a centre line, and lay the middle course first and at an angle of 45 degrees, the other courses will then follow, and the points may be kept right by means of a line drawn parallel to the centre line.

In a panel, the first brick starting from the corner should be set to a small set square, forming a right angle and two angles of 45 degrees, and measuring from the base to the apex 3 inches, or whatever the bricks will work.

DUTCH BOND.

Fig. 41 is an arrangement called Dutch bond. It is a modification of English bond, the "closer"

Fig. 41.

being omitted and a three-quarter "stretcher" used on the "quoin." In every third stretching course a "Flemish header" is introduced next to the "quoin" brick, by which means the "stretchers" in that course are pushed forward, and overlap the "stretchers" below 4½ inches, instead of being "plumb" over them as in other bonds. The advantage of this bond is that additional strength is imparted to the wall in the direction of its length, and that without diminishing its transverse strength. A writer in the *Builder*, from which Fig. 41 is taken, speaking of this subject says : " As regards construction in common English and Flemish bonds, no greater tie in the direction of the wall is obtained than 2¼ inches which one brick overlaps another. If, therefore, a fracture takes place, the crack runs down the wall, following the joint with only that small deviation from a perpendicular line; but by the Dutch method a crack would have to follow 4½ inches to the right or left in the courses containing the 'Flemish header,' or else break through the bricks. Clearly, therefore, we have some additional strength, the lap between the courses of 'stretchers' being as much as 4½ inches."

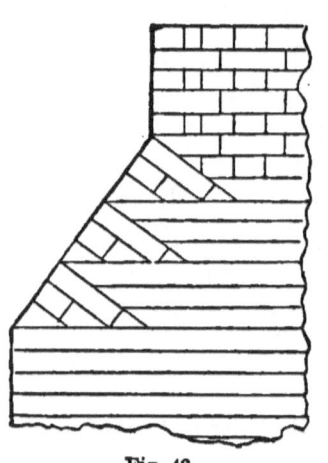

Fig. 42.

The adjoining Fig. 42 shows the way in which buttresses and chimney stacks are reduced. They are generally " tumbled in " at an angle of about 60 or 70 degrees. The beds of the bricks should always be at right angles to the " tumbling in." The bond on the " battering " jamb will be the same as on the upright jamb below.

"Keeping the Perpends."

Architects usually specify that the "perpends" shall be kept, or, in other words, the vertical joints are to fall in plumb lines from top to bottom. Owing to the difference in the sizes of bricks, this cannot be done with bricks as they come to hand; they must be sorted to a length, or cut where necessary, by the bricklayer as he proceeds with his work. This would add to the cost of the work, and, as cost has to be considered in most buildings, it is seldom done. But if the bricklayer carry up a plumb line in the middle of large piers, and work his bricks between that and the plumb reveals or jambs, he will be able to keep his "perpends" tolerably regular. The "closers" should be cut to a $2\frac{1}{4}$-inch gauge.

Toothings.

Toothings should not be allowed in a building where they can possibly be avoided; they are a source of weakness, and very often a disfigurement to a building. When building into toothings, the bricklayer seldom takes the time or trouble to make solid work; and where they have been can very often be traced in buildings that have been up but a short time by the pointing having fallen out right down the line of toothings. This is caused in frosty weather, by the expansion of moisture which has got into the hollow parts of the toothings, forcing the pointing from the brickwork, to be washed off by the first heavy rainfall. Where toothings are unavoidable, they should not be carried up in a straight line from bottom to top, as they usually are, but should be stepped back every few courses, so that the new work may be bedded solidly here and there. When building new work into old, a chase is preferable to a toothing, as the new work is left free to settle. But in a front where new work has to be built into an old toothing there should be no mortar used in the toothing; the new work should be kept a trifle high above the old, and the joints of the toothing filled in after the building is up. Among the characteristics of good brick-work are solidity, perpendicularity, smoothness; the vertical joints carry a plumb line from top to bottom; the "cross" joints of the "stretchers" fall immediately in the centre of the "headers," and the bed joints are neither too thick nor too thin.

GROUTING.

"Grouting" is the practice of using mortar or cement in a semi-liquid state to fill up the open joints in the work, the result of careless or bad workmanship. In some works every course is "grouted" in; in others every four courses.

"Grouting" is not the best way to obtain solid walls, for the mortar being in a semi-fluid state, the excess water is absorbed into the bricks of which the work is composed, and, as a consequence, the "grouting" shrinks or subsides, leaving the joints or interstices only partially filled. A better process is that of "*larrying-up*," which is, after having laid a course of bricks on each side or face of the wall, to put a proper amount of mortar in the wall, and by the addition of water, and the use of trowels, shovels, or a larry, to reduce it to such a consistency as to be able to swim in the bricks solidly. Even in this practice there is a subsidence or shrinkage of the mortar, with the same effect, though in a less degree, as described in "grouting." But the best and proper plan is undoubtedly that of putting up the joints solidly through each brick as it is laid, and having the mortar of such a consistency as to be able to draw the joints up solidly when filling in the middle of the wall.

Flues.

Of the abominations of a bad building, bad flues are second only to bad drains. The causes of smoky flues are as follows. The sectional area of the flue is either too large or too small. Its sectional area is cramped, the "cramp" generally occurring in sharp bends, close to a floor, where the bricklayer has to make room for another fireplace. The flue is too short, or is not carried up high enough to be above some adjoining building or contiguous wall. There is too much air-space below the throat of the flue, or, in bricklayers' phraseology, the wing gatherings are not brought over fast enough. In considering the scientific principle of flues, we should remember that the properties of air in their action are very similar to those of water. A stream with a straight smooth course flows swiftly and regularly, while one with a rugged winding course is full of eddies and whirls, and flows with a retarded velocity. So it is with flues. An unused flue contains a column of cold air in equilibrium with the surrounding air. This column of cold air must be rarefied or heated before a good draught can be obtained, when the denser air rushes in, pushing the lighter up. This will account for the fact that a flue never draws so well when the fire is first started as it does some little time after.

Where the flue is unnecessarily large, a larger volume of air has to be rarefied, and it also admits of a possible down draught, or in other words an ascending and a descending column, in consequence of the heated air not filling the flue. Where the flue is "cramped" somewhere in its length, the cause of smoking is that the smoke is checked in its ascent just were the "cramp" occurs, the smoke escaping with a retarded instead of an increasing velocity. Sharp bends have the same effect, though in a less degree, as "cramps." Yet it is a common thing to hear bricklayers advocating sharp bends in flues to increase their draught.

Every flue should be formed with sufficient bend to prevent the daylight and rain falling upon the fire.

Where a flue terminates below an adjoining wall, it will often smoke in consequence of a down draught, caused by the wind striking against the wall and in its rebound passing down the flue, or at least obstructing for a time the passage of the smoke from the flue, which in effect is similar to a down draught. Where the throat of the flue is formed high up above the chimney bar there is a large volume of cold air collected which has to be heated or rarefied to get a proper draught; until this takes place the smoke is obstructed in its ascent, and driven back into the room.

To cure these evils, innumerable contrivances have been invented, of various forms and different degrees of ugliness, and it is almost rare to see a house in the metropolis that is not surmounted with one or more of these articles, each advertised as a panacea for smoky flues. These so-called remedies are (with the exception of the "blower") always applied to the top of the flue, when in fact the remedy is generally required at the bottom or somewhere in the length of the flue. We would give the following advice for flue building. Form the throat of the flue as low down as possible, and let the sectional area be the same throughout its entire length, avoiding all bends beyond what is necessary to hide light. Where bends cannot be avoided let them be as easy as possible, and carry the flue well up above contiguous structures, and let it be pargetted smoothly inside. In building flues "coring holes," 12 × 14 inches, should be left out on every floor, or at least where every bend occurs, and a piece of board put in to catch the mortar and brick rubbish that fall while in erection. By

this method the flues may be easily "cored" or cleared without the aid of a chimney sweep. Flues for dwelling-houses are generally for registers, 9 × 14 inches, and for kitchens 14 × 14 inches.

Fig. 43 is the plan of a fireplace and flue for a register stove, which we insert by permission of the originator, H. Saxon Snell, Esq., F.R.I.B.A.

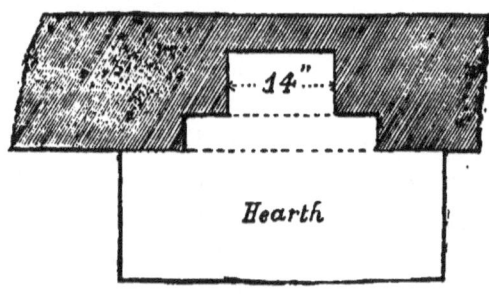

Fig. 43.

The peculiarity and advantage of this fireplace is that the sectional area or throat of the flue commences immediately on the chimney bar, doing away with the necessity of wing gatherings and the possibility of cold air collecting round the base of the flue. This for its economy of construction and efficiency of action recommends itself for general use.

All chimney stacks from the part where they pass through a roof, or from the point where they separate from a wall with which they have been in junction, to their tops, should be built in cement and sand instead of with lime mortar.

Where several flues are grouped together in

one stack, instead of dividing them with the usual 4½-inch brick "withes," Boyd's flue-plates (iron plates ⅛ inch thick, and about 12 inches square, fitting into each other with a tongued and grooved joint, and built into the sides of the stack) are often introduced to economise space.

To ensure that flues shall have the same sectional area in their entirety, they are sometimes built round a wooden section-box, open at both ends and with a wooden "strap" to take hold of, that the box may be pulled up from time to time as the work progresses. The box is placed in the space intended to be occupied by the flue, and the bricks carefully laid with full joints against the box, which is drawn up about every two or three feet. In some cases the pargetting is dispensed with, and the joints struck instead. Good flues are undoubtedly obtained in this way. The same end is obtained by the use of Doulton's terra-cotta flue-pipes; but when built in small detached piers (as they sometimes are), they prove a source of weakness by interfering with the bond of the work. Where they are grouped in stacks there should be a space of 4½ inches between each pipe, to admit of bonding the stack in the direction of its width.

SECTION II.

ARCHES IN GENERAL.

Arches.

Arches are of various kinds, but those which the bricklayer has to deal with are either circular, segmental, scheme, elliptic, or Gothic. To the young operative, and in many cases to the aged workman, they are veiled in mystery, though a little application and determination to understand them would soon make them clear to the operative who would be master of his trade. Time was when the arch-cutter would box himself up and carefully tack strips over the chinks between the boards that prying eyes might not penetrate into his cutting-shed and discover the craft by which he held himself superior to his fellow-workmen. This jealousy and exclusiveness is still alive, though it is being slowly trampled under by means of the flood of light that is spread abroad, and is still spreading, from technical classes and technical publications. If the young workman will but set to work in earnest, there is every facility to acquire technical knowledge, and to make himself, as a workman, superior to those who have gone before, and who,

> "By geometric scale,
> Did gauge the size of pots of ale."

Let him but catch that spirit breathed forth in

Longfellow's lines to Strasburg Cathedral, and success will surely be his:—

> " A great master of his craft,
> Edwin von Steinbach; but not he alone,
> For many generations labour'd with him.
> Children that came to see these saints in stone,
> As day by day out of the blocks they rose,
> Grew old and died, and still the work went on,
> And on, and on, and is not yet completed.
>   The architect
> Built his great heart into these sculptured stones,
> And with him toil'd his children, and their lives
> Were builded with his own into these walls,
> As offerings to God."

The word *arch* implies an arrangement of bricks or other material in which all its parts—we might with equal propriety say particles—are in equilibrium; or, in other words, that the pressure or thrust to which it is subjected is transmitted from one course to the other, and distributed throughout the whole of the arch, each course or voussoir taking its share. Every bricklayer who has turned an arch will have noticed that this condition is not obtained by simply turning the arch on its centre and keying it in, the tendency being for the arch, by reason of its own weight, to spread out at the springing, or if this be prevented to buckle up at the haunches, to prevent which and bring about equilibrium, calculations have to be made so as to apportion the weight at the haunches to resist or counteract the thrust from the crown. Such mathematicians as Dr. Hooke, Huygens, Leibnitz, and many others, devoted much time and attention to the solution of the principle of the arch under the name of the *catenary curve* (Latin *catena*, a chain); and the conclusion *they* arrived at was, that the true shape of an arch is that into which a chain would arrange itself if freely suspended from two points whose distance apart is equal to the span of the intended arch. We have mentioned these things because, considering the way in which arches are often *thrown* together, it is well that the artisan should know there is a principle involved in their construction.

Relieving Arches.

Relieving arches should be turned over all lintols where practicable, and should spring clear of their ends. They should not be built, as they generally are, solid on the brick "core," whereby the weight of the wall above is transmitted from the arch to the "core," from the "core" to the lintol, and from the lintol to the frame, very often to the great injury of the latter; but should be built at least $\frac{3}{4}$ inch clear of the "core." This can be done by putting a layer of sand $\frac{3}{4}$ inch thick on the core, and raking it out with a trowel or piece of hoop iron when the arch is turned, that it may take its own bearing. They should be turned in compo.

The above remarks apply to where the window and door frames are built into the brickwork during erection; and more particularly to arches intended to relieve free-stone rectangular door and window heads. It is not an uncommon thing to see such heads fractured right through their depth in about the middle of the openings which they span, and kept from falling only by the weight of brickwork upon their ends; though the architect has been careful to provide against superincumbent weight by the use of relieving arches, but which, through inexperience or want of judgment, or some other cause, have been built upon a solid "core."

Plain Arches.

All arches put in with bricks as they come from the brickfield come under the term plain arches, and are built in concentric rings of $4\frac{1}{2}$ inches laid as "headers" on edge, instead of bonding by "stretchers," to avoid the large joints that would unavoidably occur at the extrados, thereby decreasing the strength of the arch unless it were built with cement, or a strong hydraulic mortar, as lias.

The Skew or Oblique Arch.

This arch is used in the construction of bridges over roads or waterways where the bridge is not at right angles to the road passing under it.

Two very remarkable arches of this kind may be seen on the Metropolitan District Railway at Brondesbury, and which the writer believes to be the only bridges so constructed. Of these we will speak hereafter.

To set out and understand drawings of the skew arch, a knowledge of solid or descriptive geometry is indispensable; but as the setting out is generally performed by the engineer or inspector of works, we will confine our remarks to that portion of the work which properly belongs to the operative bricklayer. A B C D, Fig. 44,

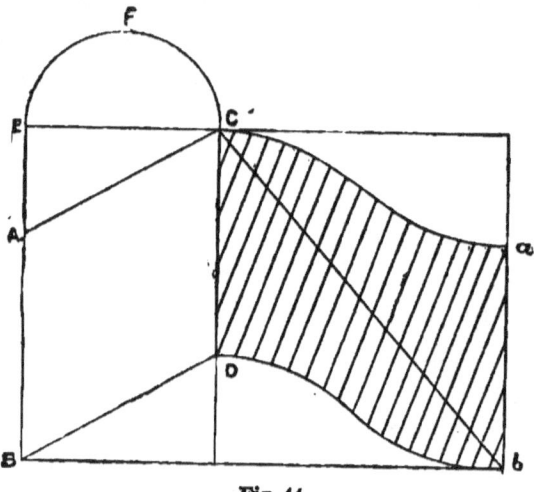

Fig. 44.

represents the plan of a skew arch of which E F C would be a section cut square with the abutments. E C A is called the angle of skew, for it shows how much out of square the face of the arch is with the road. *a c* is the face of the arch, and as the "bed" joints (called by engineers "coursing" joints) start square from the face, they must run in a diagonal direction across the centre, as seen in C D *a b*, which is a development of the soffit of the arch. To make this clear, we will suppose the courses to be pencilled on the centre, and a sheet of white paper folded round the centre and rubbed until the pencil marks be transferred to the paper. If the paper—

ARCHES IN GENERAL. 51

fastened at C D, the abutment line—be now unfolded from the centre and spread out on a level surface as in Fig. 44, we shall have a development of the soffit of the arch. c *a* is the length of the line on the centre from c to A. D *b* is the length of the line on the centre from D B, and is

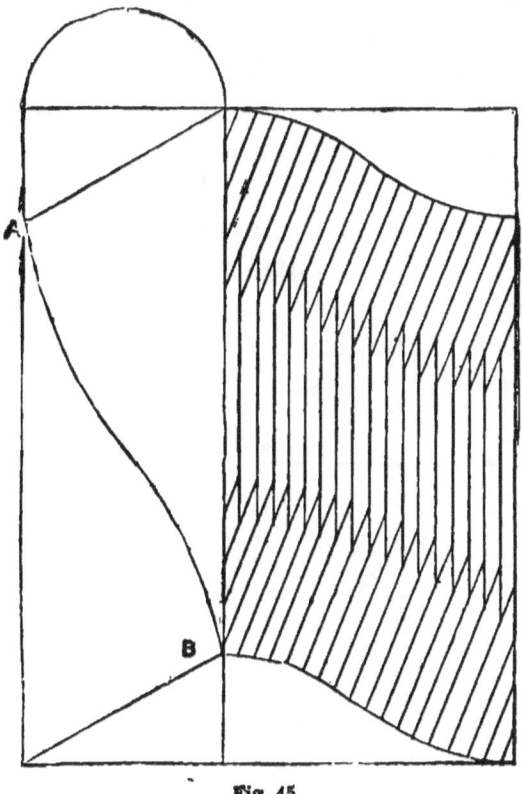

Fig. 45.

parallel with c *a*. c *b* is the length of a line on the centre from c to B.

In long skew arches the bricks, instead of being laid on the skew all through the arch, are

D 2

arranged as in Fig. 45, where the skew courses are intersected by courses laid parallel with the abutments. The skew courses are marked on the centre by means of a "coursing mould," which should be supplied by the engineer or inspector in charge of the work. A B is the plan of a line on the centre from A to B. All the courses on the centre will be so many spirals or screws parallel to each other. Each brick on the face of the arch will require a different bevel, but by far the easiest and the best way to get these will be to let the bricks stand well out in front of the face line, and cut them off to the line of work when the centre is struck. But when the bricks used are too hard to be cut, such as Staffordshire blue bricks, they must be moulded to the required bevels.

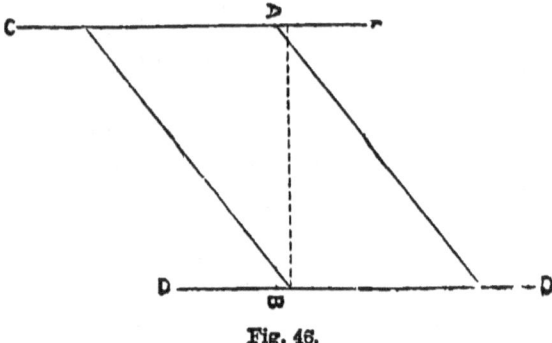

Fig. 46.

SKEW ARCH AT BRONDESBURY.

The remarkableness of this arch or skew is not alone in its construction, but in the angle that it makes with the roadway that it spans, the angle

ARCHES IN GENERAL. 53

being so acute as to cause the abutment line or skew-back of one side to fall *without* the abutment line of the other side. This is shown by the line A B at right angles to C C, D D, Fig. 46.

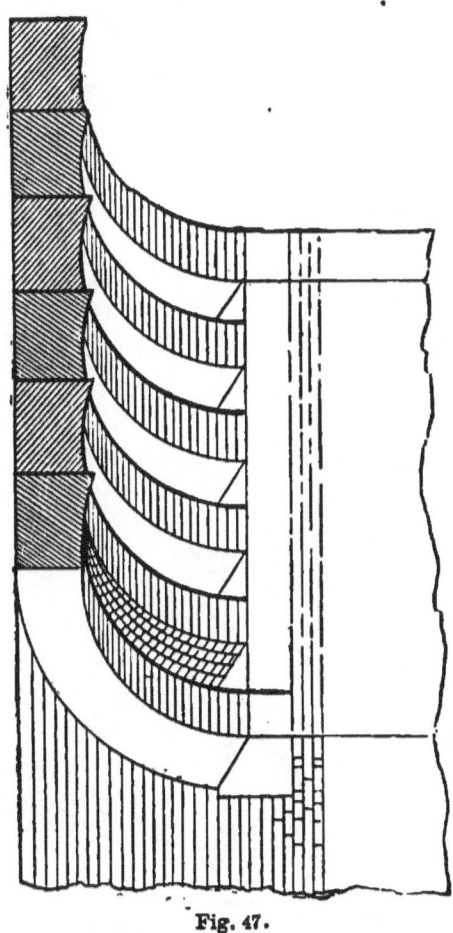

Fig. 47.

Let us imagine that across a given road we have to construct a bridge whose angle of skew

54   BRICKWORK.

shall be equal to that on the accompanying

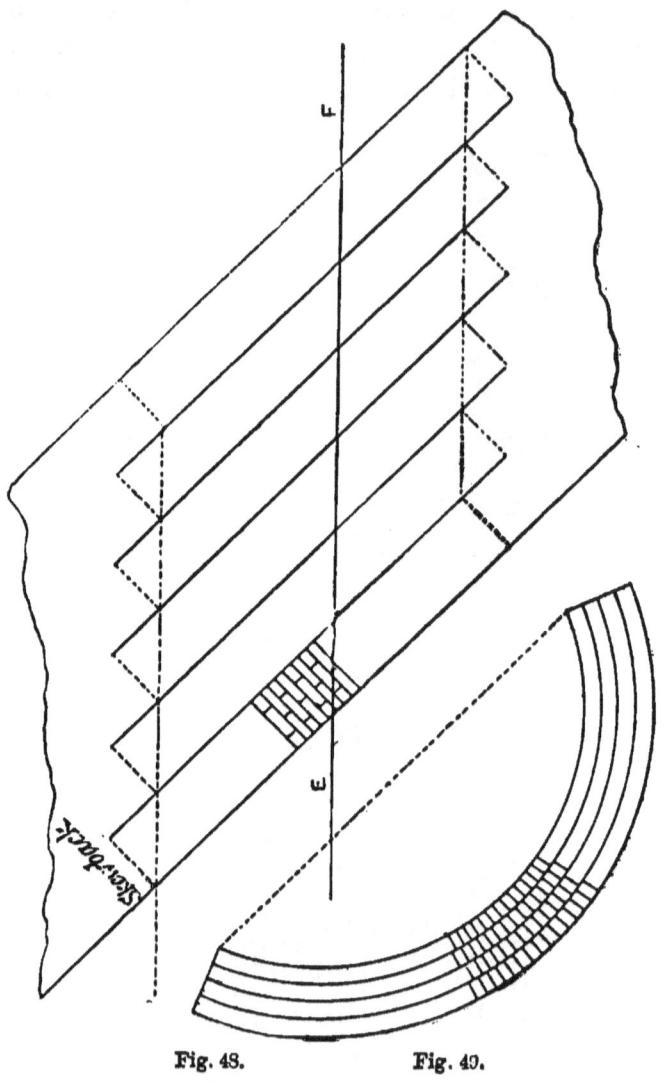

Fig. 48.   Fig. 49.

Fig. 46. It is clear that we cannot construct it on the principle of the ordinary skew arch, viz.

to take the courses (starting square with the face of the arch) as so many spirals across the centre, finding their abutment in the line c c, and as we have explained at page 50. An arch so constructed could not stand, for the lines of force, or thrust, acting at right angles to the abutments, would find no resistance, and consequently collapse. But the engineer who designed the bridge in question, seeing this, fell back on the principle that should regulate the construction of all arches where strength is required, that *the bed joints shall be in the line of radii, and on the soffit parallel with the abutment*, and thus in the simplest, yet most effective manner, solved the otherwise difficult problem. Fig. 47 shows the sectional elevation on the line E F in plan, Fig. 48 the plan, and Fig. 49 the face arch in elevation of a bridge somewhat similar to that at Brondesbury, constructed in the same way, and involving the same principles.

The plan of the abutments and skew-backs are shown by dotted line (Fig. 48).

The following are approximate dimensions of this bridge, which we have taken by step measurement : distance between abutments, 45 feet ; depth of bridge, measured along the abutment, 26 feet ; rise of arch from cord line to crown of soffit, 20 feet ; projection of one abutment beyond the other (D beyond A, for instance, Fig. 46), 36 feet. The arch is made up of twelve 4½-inch concentric rings of brickwork.

Water Conduit.

Fig. 50, a section of a water conduit in Massachusetts, U.S.A., upon which the author was engaged as inspector of works, is worthy of notice, as showing the construction resorted to where a bad bottom occurs. In this case a large

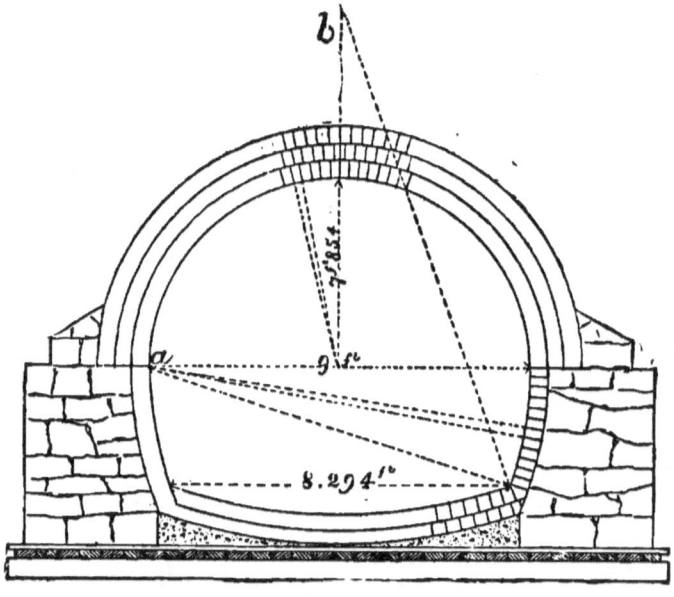

Fig. 50.

portion of the work (which was eighteen miles in length) ran through very swampy ground, a natural watercourse that drained a large tract of the adjacent country, and at times so great was the pressure of the water as to cause it to rise in a natural fountain 6 or 7 feet above the excavations. When this occurred, stones of sizes similar to those of which the retaining walls were built were shot into the hole until the water subsided or found an easier outlet elsewhere. It was also necessary to keep pumps working night and day.

The bottom consisted of 6 × 6 inch transoms, 18 inches apart, to which were spiked 2-inch planks, and these in turn were covered with 1-inch boards, with joints properly broken, as in flooring. The invert, when the side walls were built, was formed with concrete ready to receive the brickwork. The whole of the work, including concrete, was built in Rosendale cement, manufactured in Rosendale, New York, from a stone found in that locality, which when manufactured is in colour very similar to Roman cement, but less quick in setting, and attaining a greater ultimate strength. It will be noticed that the sides of the invert are struck from the springing line a, and the bottom from b, and that to get the requisite skew-back for the top and bottom beds, a *purpose made* brick is introduced, whose beds are in the line of radii from a and b.

Sewers are constructed on the same principle as water conduits, with this difference, that while strength and sound work suffice for the latter, to these must be added smoothness for sewers, avoiding all "shoulders," "lips," protuberances, or other irregularities likely to increase friction, or in any way retard the velocity of the sewage. Where the flow is intermittent they are generally built egg-shaped, to minimise the frictional area.

Groined Vaulting.

Brick groin-vaulting (a very neat sample of which may be seen at the entrance to Winchester Flats, Winchester Terrace, Chelsea Embankment) was at one time very much in practice, but moulded stone ribs finishing at the apex with a carved boss now generally take the places of the brick groins. Samples of this kind of work may be seen at St. Augustine's, Kilburn; St. John's, Auckland Road, Upper Norwood, and the red brick church adjoining the Croydon railway station, all designed in that style known as the thirteenth century, or Early English, by John L. Pearson, R.A. Some good Gothic vaulting in red brickwork may also be seen at the New Law Courts, London. In executing the groin the bricks must be cut so as to form a return on the intersecting arch or vault; but a proper bond, as in square angles, cannot always be obtained, for, instead of the bricks returning from right to left and from left to right every other course, it will be found necessary to sometimes return several courses in succession, all from one side, before getting what bricklayers would call "a tie." This is caused by the groin not getting away fast enough from an imaginary line drawn across the arch from E to G Fig. 51. It is also impossible to keep the perpends regular near the groin, but they should be kept as regular as practicable with a good bond on the groin.

Before the bricklayer can cut his bricks, the

centres must be placed in position, and the bricks can then be cut to fit the intersection, which they should very accurately, and when the centres are "struck" present clean and well-defined arrises. Fig. 51 is the plan of two semi-cylindrical vaults, intersecting in the groins E F

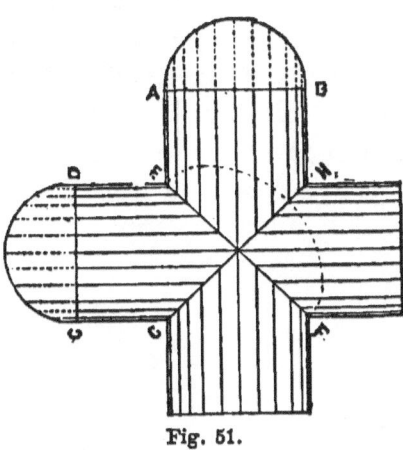

Fig. 51.

and G H. The curve formed by the groin is an ellipse shown in angular elevation on E F by dotted curve. Sections of the vaults are shown on A B and C D. Sometimes instead of being as here shown, the intersecting arches are Gothics, or one Gothic and the other semi-cylindrical; but if what we have written be understood no difficulty will present itself. In all such cases the bricklayer must space his centre out into courses, and turn the arches as any other arch, with the exception of the groin, which must be treated as described.

In Gothic vaulting, as described above, in

which the spaces between the stone springers are filled in with brickwork, the setting out of the courses is done by marking upward from the intersection, or springing of the ribs, an equal distance along the cross rib and the diagonal or converging rib, and connecting these two points with a line. Upon another line at right angles with this, the courses may be pricked in from springing to apex, and their beds shown by lines parallel with the first line, connecting the ribs. A sample of fan-groining, in red brickwork, may be seen at the subway to the Crystal Palace, Sydenham.

SECTION III.

GAUGED-WORK AND ARCH-CUTTING.

Gauged Work.

"Cutting" is divided into "axed work" and "gauged work." In the former the bricks are finished with the Scotch, with just a rub or two round the rubbing stone to take off the irregularities of the beds, allowing $\frac{3}{16}$ of an inch joint for tuck-pointing. This work is intended to represent "gauged work," and is supposed to be a trifle cheaper. "Gauged work" is a very superior kind of brickwork, executed in soft bricks set with a white putty joint, which should not exceed the thickness of a new sixpence. The bricks used are Fareham rubbers and T. L. B. rubbers for red work; and malm-cutters and sometimes white Suffolks for malm or stock work. Of red bricks Fareham Rubbers are the best; they are of a close, firm texture, will carry a sharp arris, and weather well; in colour they are cherry red. No. *ones* T. L. B.'s are good bricks, though less firm than Farehams, but of an even texture; they are divided by colour into two classes—cherry-red and orange tint. The orange is generally used, as they contrast well with the red building bricks, but will not carry so sharp an arris or weather so well as the darker bricks.

"Gauged work" is often objected to on the ground that it will not resist the action of the weather. This we can refute by our own experience, for we have taken out old "gauged" arches in malms that have withstood for forty years the acids contained in London smoke, and have shown no signs of decay or disintegration. We can cite another instance of the indurating properties of "gauged work" in white Suffolks when exposed to the action of the atmosphere. During the erection of the Rackham Street Marylebone Infirmary, some geometrical windows in these bricks had to be cleaned down some three or four months after erection. This process had to be done by rasping the face of the brickwork, and so hard had become the bricks that it was with difficulty that an impression could be made at all, the rasps sliding off the work and leaving a black mark! Bricks in this condition are said by bricklayers to be case-hardened.

This so-called case-hardening we attribute to the process of setting. In good setting the bricks are always soaked (not to saturation) in water, which in a building in course of erection always contains more or less lime in solution, which is taken up by the brick while soaking, and by exposure to the atmosphere becomes carbonised and forms a hard coating, as it were, upon the face of the brick. This case-hardening is also attributed to "the silicic acid in the clay acting upon the chalk so as to form some of it into a silicate of lime." Rubbers are purposely made much larger than the ordinary building bricks to allow for cutting and gauging them four courses to the foot, though as a rule they will not hold out or bed more than 11½ inches with close joints. T. L. B.'s as they come from the brickfield measure $10\frac{1}{2} \times 4\frac{7}{8} \times 3\frac{1}{8}$ inches.

They are also obtainable 12 inches long, but bricks this length are only required for Camber arches, or Gothic arches whose bed joints radiate from the centre, as in Figs. 57 and 58, in which so much of the brick is cut away to form the long bevels on the soffit and crown, that the ordinary sized bricks will not "hold out" to the required lengths, and have therefore to be lengthened, where necessary, by forming the long "stretchers" out of two three-quarter bricks (this will be best understood by examining a few actual camber arches); to obviate which, the 12 inch bricks are made.

Setting.

In setting "gauged work" the joint is taken up by absorption by holding the bed of the brick in contact with the putty, which must have the proper consistency and be kept in a small putty-box made with a level top, so that the setter can rest or steady his arm upon it while "dipping" his brick. Before putting the brick in place, the putty is scraped off the middle of the "bed," that it may set or joint more evenly. The joint should not be touched after the brick is "bedded," but should be left full like a small bead. Stone lime should be used for setting, as chalk lime is not fit for out-door work. Axed-work is generally set with putty and cement. If the work has to be carved deeply, it is best to build it all "headers," and "grout" it in solidly at back with Portland cement, that the bricks may not break up or get disturbed under the chisel of the carver.

A composition of whitening and patent knotting is more frequently used than lime-putty for bedding or setting work intended to be carved, and for ornamental key-blocks made up of two or more bricks. It will be found most convenient to put such keys or blocks together in the cutting-shed, and take them upon the building to be set as one piece of work. These remarks apply equally well to the niche hood in every particular. Gauged work intended to be bedded in the above composition should be quite free from moisture; but the bricks should not be placed round a fire for this purpose, as they often are, for by so doing they are made fragile and are easily broken. It is, therefore, very imperative that a good water-tight cutting-shed be made for the bricklayer and another shed for the bricks.

Drawing and Cutting Arches.

This forms a very important branch in the trade of the bricklayer, and a thorough knowledge of it is indispensable to the operative who would be master of his trade. In this section we will endeavour to make clear not only the setting out of the various arches, but how to take off the bevels and moulds, and apply them to arch-cutting.

An understanding of this will not be so difficult as may at first sight appear. The tools required

GAUGED-WORK AND ARCH-CUTTING.   65

for this work are—the rubbing-stone (which should not exceed in diameter 14 inches), hammer, boaster, Scotch, scriber, and tin-saw. The scriber is a small tin saw, used for marking the beds and bevels on the bricks.

THE BULLS-EYE

Should have four keys, *a, b, c, d,* which when possible should be "stretchers;" but as this cannot always be done unless by very much reducing the size of the courses (technically called *voussoirs*), they are, therefore, frequently put in as in Fig. 52. The face mould for this arch is obtained by making a wooden

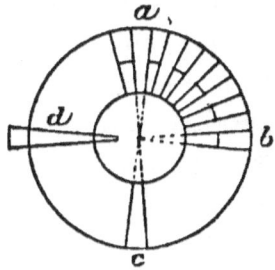

Fig. 52.

pattern, as at *d*, on which the actual length of the brick is marked, and also its bevel, which is taken off the drawing by placing the stock of the bevel along the bed joint, and moving the blade until it coincides or is in line with the soffit of that particular brick whose bevel is required. All the courses have the same bevel and the same length. It is usual to have two moulds made, so as to trace or traverse the courses round the arch, to ensure that the key brick will come in rightly (though one mould and two parallel straight edges would do equally as well); for if the mould be in the least inaccurate, the inaccuracy will be transmitted to each brick, and this multiplied by the number of courses in the arch (in this case 36), supposing the inaccuracy to be $\frac{1}{16}$ of an inch, would amount to $2\frac{1}{4}$ inches, in all probability the thickness of a course. Having proved the moulds, the pattern brick or soffit is marked lower down on the mould, that the brick when cut will be the thickness of a joint less than the brick shown on the setting out. The bevel of the thick end or *extrados*, as it is named, is the same as that of the soffit.

The arch cutter will find it most convenient to have a square piece of wood, $4\frac{1}{2}$ by 9 inches, with parallel sides, which held flush with the soffit will give the exact place and bevel of the cross joint, and held longwise the length of the brick and its end bevel.

In cutting, the first operation is to square the bed and face of the brick, after which the soffit is bevelled. The brick is then placed on a bedding board (a piece of slate or wood with a straight even surface) in the same position that it will have in the arch. The face mould is applied to the brick with the soffit mark against the soffit of the brick, and the scriber drawn along the top edge of the mould marks the wedge shape which the brick will have when finished. The back of the brick is marked in the same way, and is then finished with the boaster, Scotch, and rubbing stone.

Semi and Segmental Arches.

What has been said of the bulls-eye applies in every respect to the semi (Fig. 70) and the segment arch. To draw the curve (Fig. 53), the span and rise being given, bisect the line *a b* with *c d*; join *e b*, and bisect this line with *i h*; a line drawn from *i* through *b* will give the line of skewback. Taking the distance *i b* in the compass, with one leg

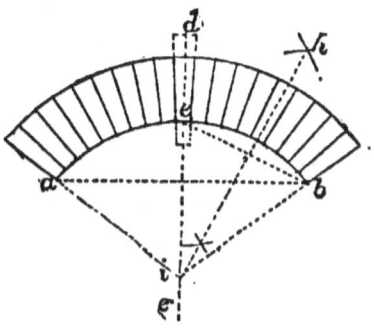

Fig. 53.

fixed at *i*, the lower curve may be drawn from *b* to *a*. Nine inches measured along the skewback from *b* will give the point from which to draw the outer curve. On the outer curve, with *c d* as centre line, set out 3 inches, or whatever a brick with its joint will hold out, and with the mould (shown by dotted lines) trace the courses down to the skewback, increasing or diminishing the thickness of the brick as may be required by raising or lowering the mould.

THE CAMBER ARCH.

Fig. 54 is a camber, 12 inches deep, in Flemish bond. The skewback is obtained by taking in the compass the distance *a b*, and from these points, with *a b* as radius, drawing the inverted Gothic; a line from *c* through *b* will be the line of skewback, or springing. To draw this arch when the skewback is given—say 4½ inches—from the centre line set off the distance between the reveals from *a b*; 12 inches above the springing, draw the line $d\ e$, and from centre line along $d\ e$ measure off a distance $4\frac{1}{2}$ inches beyond the reveal; from this point draw a line through b, intersecting the central line in c. On $d\ e$ measure off $1\frac{1}{2}$ inch each side of the centre line, or whatever a brick with its joint will measure. Lines drawn from these two points to e will represent the key, and also the face mould. Make two moulds 9 inches

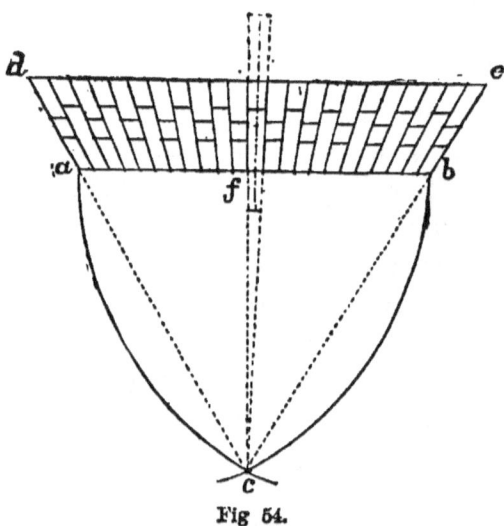

Fig 54.

($4\frac{1}{2}$ inches at each end) longer than the key. With the mould, shown by dotted lines, upon the key, on one of its edges, f, where $a\ b$ meets it, make a pencil mark. Put the other mould on top of this and transfer the mark to it. With the two moulds, keeping the pencil mark always on the line $a\ b$, traverse the courses in down to the skewback as described in the bulls-eye. Take off the bevels, starting from the skewback, and pencil them upon the mould, 1, 2, 3, and so on, as shown in Fig. 55, *a*, which is a mould with the lengths and bevels of each course upon it. One-half only of the arch need be set out. The cross joints may be cut in the courses with the saw and parallel board, as previously described, always working from the soffit. For greater accuracy and distinctness, the bevels may be pencilled on the back of the mould, at the top end, keeping them some little distance apart, and numbering them as already described. The courses may be traversed in by working from the top line *d e*, instead of from the soffit, marking on the mould,

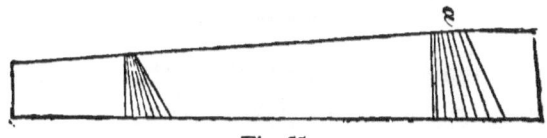

Fig. 55.

downward from the top mark, the length of each course. Having thoroughly understood the setting out and cutting of this arch, no difficulty will be experienced with any of the ordinary arches.

The soffit generally cambers $\frac{1}{8}$ of an inch to the foot.

The camber is not suited for large openings, or where any considerable weight has to be carried, as it is in reality not an arch at all, but simply an arrangement or *scheme*.

THE GOTHIC ARCH.

Bisect the line *a b*, Fig. 56, with *c d*, and draw *a d*; from these two points with the compass opened to more than half their distance draw the arcs *s f*. Through their intersections draw a line meeting *a b* in *g*, from which point with the

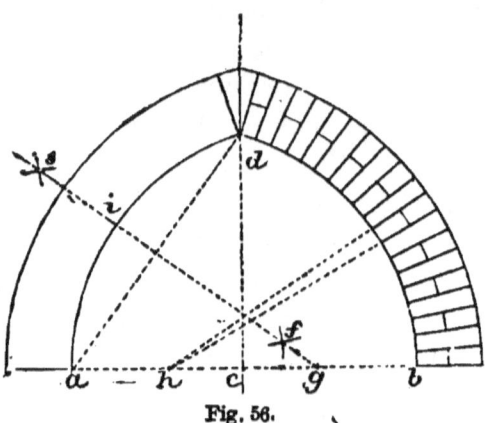

Fig. 56.

compass opened to *a*, draw the curve *a i d*, and by extending the compass, its parallel curve. From *h* draw the curves on the right-hand side. The bed joints radiate from *h* and *g*, as shown by

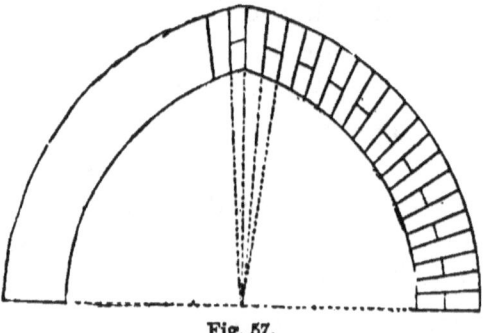

Fig. 57.

dotted lines. To do away with the very wedge-shaped key, the joints are sometimes radiated from the centre, as in Fig. 57. This key is also objected to by some on account of the oddness of its appearance at the key—a "stretcher" on one side and two "headers" on the other (this is what bricklayers call keying in with a joint), to prevent which a "birdsmouthed" key is used, Fig. 58. In the last arrangement the arch has an odd number of bricks, in the two former an even number. Whatever objections may be urged

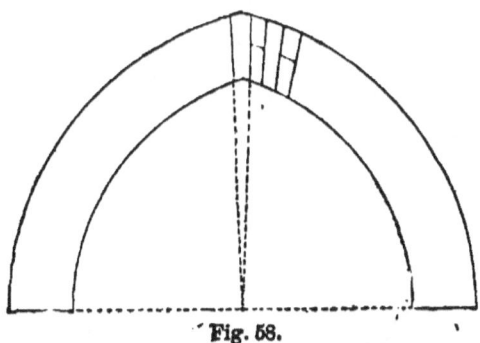

Fig. 58.

against the appearance of Figs. 56 and 57, the birdsmouthed key in Fig. 58 is decidedly wrong:—

"The essential character of the Gothic arch is derived from the absence of the key-stone, and from the presence of the perpendicular joint or opening in the centre where the archivolts rest against each other. Until we find this feature, Gothic architecture does not exist."—*Normandy: Architecture of the Middle Ages.*

Fig. 56 is made up of two segments of a circle, and the mould is obtained in the same way as that for the segment. The moulds for Figs. 57 and 58 are obtained in the same way as that for the camber, the bricks being all of a different bevel and length. These like the camber are *schemes*, not arches, as the bed joints do not fall within the *lines of radii*.

THE ELLIPSE GOTHIC ARCH.

Divide the span *a b*, Fig. 59, into three equal parts; take two parts in the compass, and with one leg fixed at *a* draw the arc *d e*, and from *d*

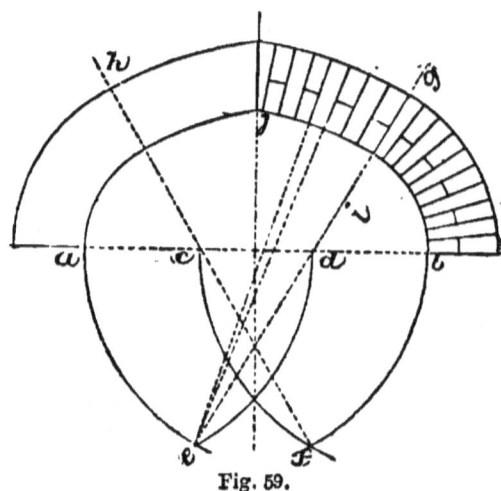

Fig. 59.

the arc *a e*. In the same way draw the arcs *b f*, *c f*. Through *e* and *d* draw the line *e g*; through *c f* the line *f h*. With *d* as centre, radius *d b*, draw the arc *b i*, and from *e*, radius *e i*, the arc *i j*. The points from which the joints radiate are shown by dotted lines. Two different face moulds are required for this arch.

THE SEMI-ELLIPSE ARCH.

Divide the span *a b*, Fig. 60, into two equal parts, *a c, c b*, and *a c*, into six equal parts, 1, 2,

3, 4, &c. From *c* towards *b* measure off two of those parts, and with the distance 4 *d* in the compass, and one leg fixed at 4, draw an arc cutting the centre line in *e*. Through *e d* draw the line *e f;* with *d* as centre, radius *d b*, draw the arc *b g*, and from *e* with radius *e g*, the arc *g h*. Two ways are here shown of putting in the courses—one in which the joints radiate from their centres

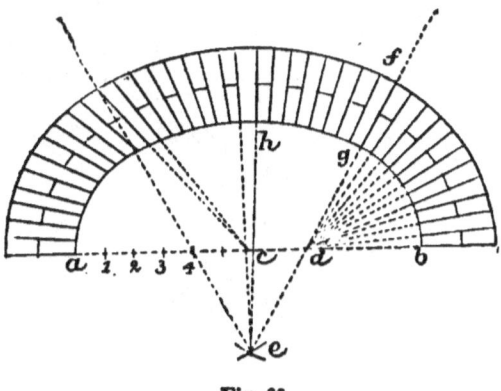

Fig. 60.

or foci *d e*, the other from *c* the centre of the opening. In the second method the lengths and bevels of each brick would be different. The first is an *arch*, the second a *scheme*, and is never adopted except in face work when, in the opinion of some people, it is desirable to have the courses all one thickness, even at the loss of strength. In the second method the mould, lengths, and bevels are taken off in the same way as those of the camber.

The Venetian Arch.

This so-called arch, Fig. 61, is made up of the camber and semi, and was a few years ago very much used in the construction of three-light

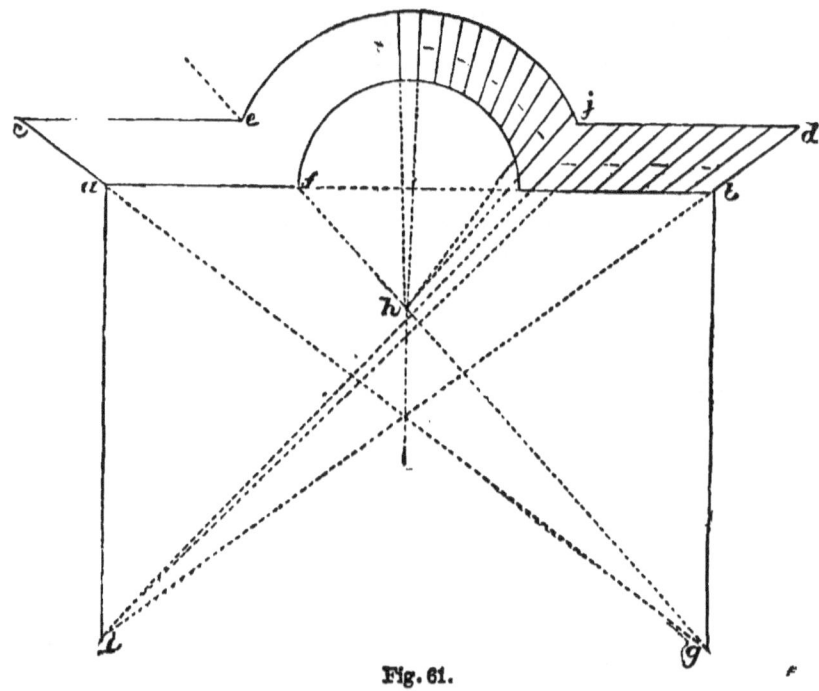

Fig. 61.

windows, sometimes with and sometimes *without* supporting mullions. Without mullions it is a very weak construction, and incapable of carrying much weight. But in this case it is generally allowed to have a bearing on the head of the solid window frame by showing less than $4\frac{1}{2}$ inches on the soffit. It is sometimes relieved by a gauged discharging arch above it. Having drawn the semi, draw the parallel lines *a b, e d*, and through their points of intersection *e f* the line *c g*. A line from *g* through *a* will be the line of skewback. This repeated on the opposite side will find *i*. Next draw the angle brick *j*, the joints in the semi radiating from *h*, and the joints in the camber from *i*. Two different face moulds are required, which with the lengths and bevels of the courses must be taken off in the same way as described in the camber.

THE SCHEME ARCH.

Fig. 62 is the same as the segment, with this difference, that instead of springing from its

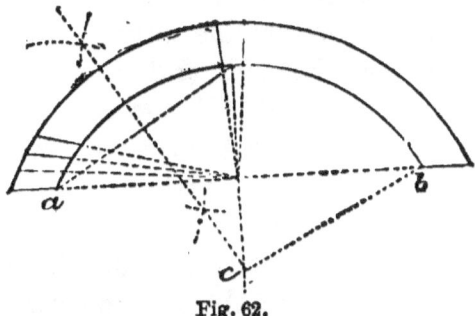

Fig. 62.

proper skewback *c b*, and its courses radiating from *c*, the curve is brought down to a level line or very near it, and the joints radiated from the centre of the opening in the level line. The *scheme* is the offspring of an antiquated and bad taste, and is not much used in the present day. One would think that its ugliness and want of truth would entirely forbid its use. It is treated by the cutter in the same way as the camber arch.

THE SEMI-GOTHIC ARCH.

To draw the semi-Gothic, Fig. 63, bisect (divide into two equal parts) the line *a b* with the perpendicular *c d*, and having determined the height of the apex *d*, from *d* draw the line *d b*, and from these two points the arcs through which the line *e f* passes, intersecting the cord *a b* in *e*.

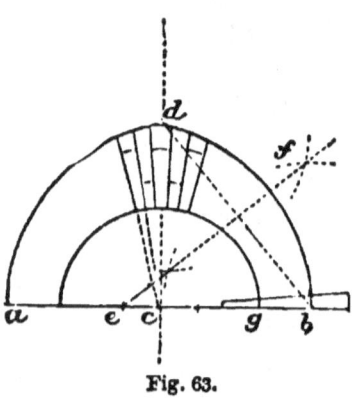

Fig. 63.

Now with the distance *e b* in the compass draw the Gothic or outside curve. Repeat this operation on the other side and the outline of the arch will be drawn. To fill in the courses divide the soffit or semi into equal parts, whatever a brick will work or "hold out," and from the centre *c* through these parts radiate the courses as shown. The moulds are taken off as described in the bullseye, and traversed from the key downward to the springing, taking care that the soffit mark on the mould always comes on the soffit of the arch. Having done this, mark on the mould the length of each course, which will also give the bevels of the top ends of the courses. The mould is shown on the springing course with the length and the outside bevel marked on it; *g* is the soffit mark to cut to; allowance must be made for the joint.

Gothic on Circle Arch.

Fig. 64 shows the way to set out the moulds for a Gothic arch in a turret or bay that is circular in plan. Draw the elevation of the arch and the plan of the wall. A little considera-

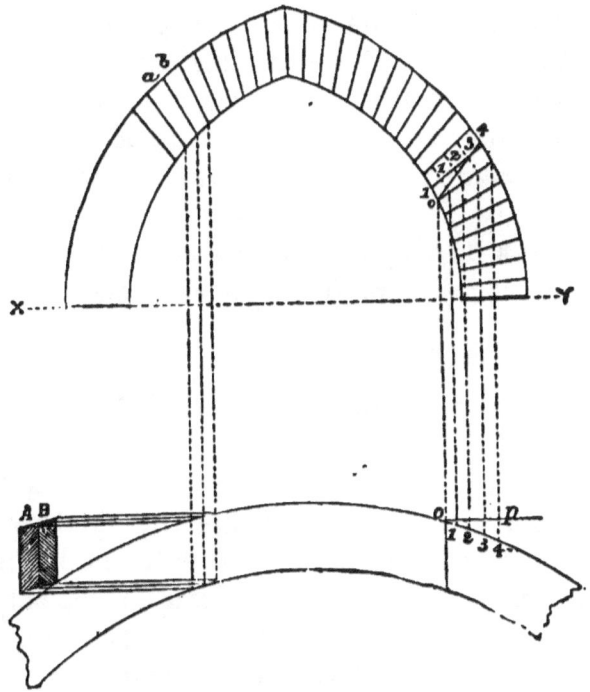

Fig. 64.

tion will show that the face of each course has a different curvature or "sweep," that at the springing having the greatest—equal to the wall itself —and the key the least, the curvature becoming less as the courses approach towards an upright position. A separate section mould must therefore be obtained for each course. Divide the bed joint of the course *o'* whose curvature is required into a number of equal parts, from which drop lines square with *x y*, and intersecting the outside curve in *o*, 1, 2, 3, 4 in plan. Draw *o p* parallel with *x y*, and transfer the distances 1, 2, 3, 4 from *o p* in plan to lines or ordinates square with the bed joint of the course whose curvature we are obtaining. A line drawn through these points will be the curvature of the section or soffit mould. By the same method the curvature of each course may be obtained. If all the soffit moulds were drawn connectedly, as A B, we should have what would be called a development of the soffit. The Gothic on circle is the same principle as circle on circle.

To Find the Soffit Mould.

From *a* drop down the two left-hand lines passing through the circular wall below *x y*. From their intersection with the two curves draw lines parallel with *x y*. Take the thickness of the soffit in the compasses, and with one leg fixed anywhere in the upper line draw an arc cutting the lower line; these four points connected will give the soffit mould A. Moulds for two course, *a* and *b*, are shown; the others are obtained in the same way. This arch in practice is generally cut by rule of thumb, or what workmen call "near enough," and rubbed down to a suitable shape when the building is up, and its faults hidden with stopping of the colour of the bricks. But where perfect accuracy is required the moulds must be obtained as shown.

SECTION IV.

ORNAMENTAL BRICKWORK.

Ornamental brickwork in this country has reached its greatest height in connection with the Queen Anne style of architecture, as elaborated in the present day. The oriel windows of the Tudor, the ornamental gables and picturesque chimneys of the Elizabethan, are all merged into it, and with such a profusion of carving as to be unprecedented in any former age. Indeed, to such an extent is this being carried as to call forth from one of our most popular architects the assertion that we are fast departing from the vernacular of our street architecture. Let us rather say, if we may use the expression, that we have entered into the Augustan age of brickwork, in which the stuccoed front with its hidden carcass of "shuffs" and "place bricks"—often the refuse of the brick-field—is superseded by that which is what it appears to be, bearing on its face the unmistakable stamp of truth!

The Niche.

Figs. 65, 66, and 67 are the elevation plan and section of a niche in Flemish bond. This is considered by bricklayers to be one of the most artistic pieces of work in connection with their trade. There are two kinds of niches, the semi and the elliptic. In the former it is circular in plan and elevation, in the latter it is elliptic in plan and circular in elevation. If that in our illustration be understood, no difficulty will be experienced with

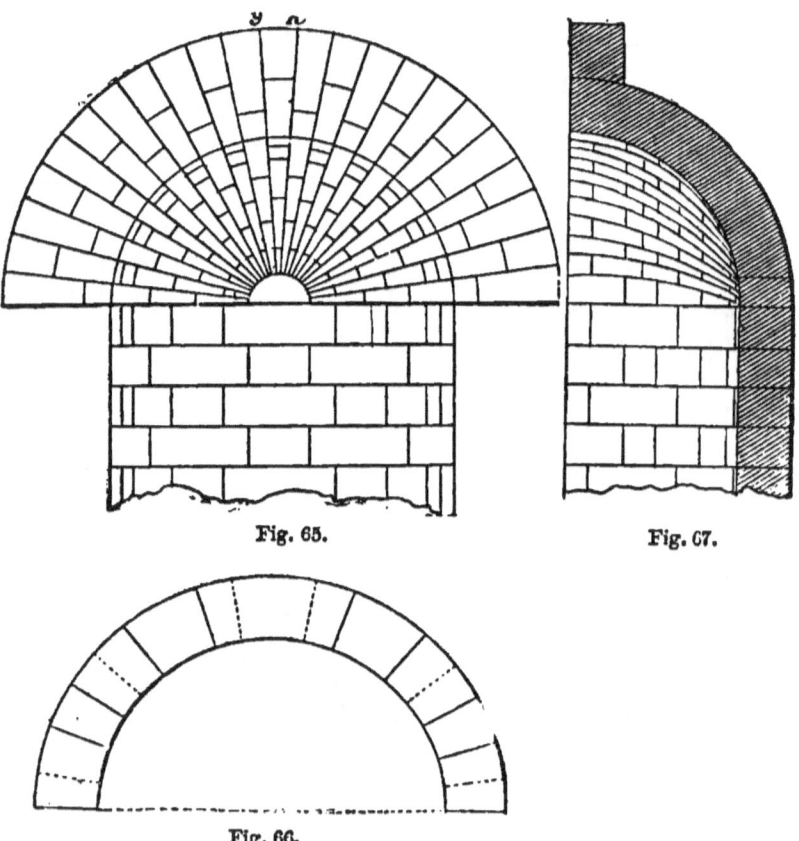

Fig. 65.   Fig. 67.

Fig. 66.

the others. The back or upright part is built to a template forming a semicircle, and the bond set out as shown on plan Fig. 66, the joints of one course being shown by thick lines, and those of the course below by dotted lines. But it is the hood, the more difficult part, that we wish to explain. To make the centre, two pieces of wood, each a semi of the same circle as the niche, are nailed together with brackets in the internal angle (Fig. 68), and the space between the brackets filled in with core, pieces of bricks and mortar, and the surface finished with plaster of Paris, by means of a template a little more than a quarter of a circle (called the generating circle) fixed with a gimlet to the back of the bottom semi. The

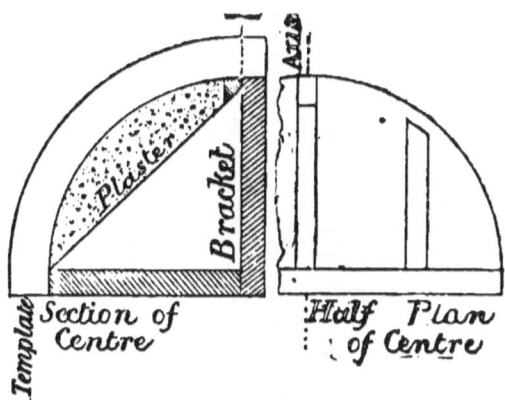

Fig. 68.

template rotating upon the gimlet as an axis, with the other end of it carried round the edge of the upright semi, a quarter of a sphere will be described or generated.

We have now got the centre or turning piece. Next draw the front arch as an ordinary semi arch, and mark the same number of courses on the top of the centre to represent the soffits. Then with a plianth straight-edge or the rotating template, mark the courses on the plaster centre, all meeting in a needle-point where the gimlet entered; but as the bricks cannot be so finely cut, a small semicircle or "boss" is introduced of such a size that the bricks at the points where they meet it will be in thickness about half an inch. The courses are all of the same length and bevel, and the soffits must be bevelled in the same way as those

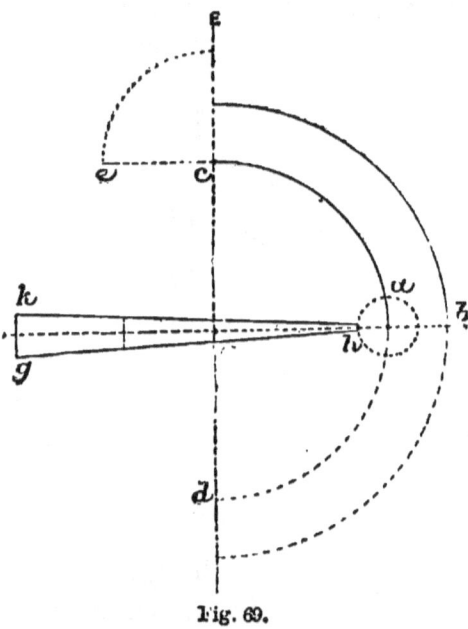

Fig. 69.

of an ordinary semi arch; and by looking at the elevation and section we see that the hood is made up of a series of semies increasing in size from the "boss" to the face arch.

The Niche Mould.

The length of the course must be measured from where it meets the "boss" to the outside of the 9-inch face arch. From h, Fig. 69, draw a line square with $c\,d$, and on it mark a distance $f\,h$ equal to the arc $a\,c$, and from f a distance $f\,g$ equal to $c\,e$, making $g\,k$ equal to $g'\,k'$ in elevation (Fig. 65); connecting these two points with the circle h we obtain the mould. The length of $c\,a$ is obtained by dividing it into small spaces and transfering them along the line $h\,f$; $f\,g$ is the length of the key brick, and is shown turned up into its proper position C E.

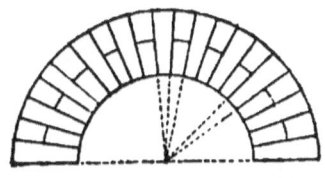

Fig. 70.

Moulded Courses.

It is the work of the bricklayer to cut and form all kinds of mouldings, dentils, entasis columns, flutings, and such like members in gauged work, leaving the more intricate, such as design and foliage, to be executed by the carver. Fig. 71 shows the kind of box that is used for cutting moulded bricks to any required section—in this case an ogee. The box is generally made to hold two headers or one stretcher. The brick or bricks, having been squared and rubbed down to the required thickness, are placed in this box and with the bow-saw roughly cut out, and then rubbed down to the section of the box with a rasp, and sometimes a piece of straight gas-pipe to form the hollow members, the bricks being

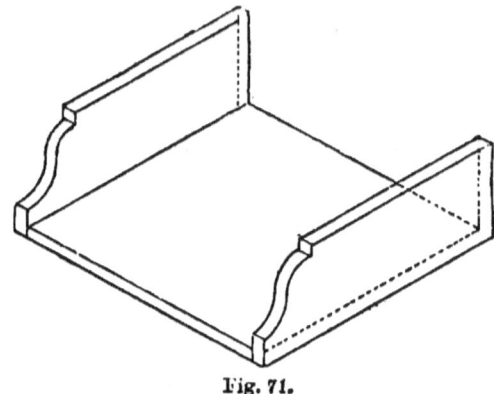

Fig. 71.

very soft. Care must be taken that the bricks be not wedged up or cramped too tightly in the box so as to "flush" the edges; and here we might mention that it is sometimes advisable to work the bricks a little wide, that in case of "flushing" they may be brought up to an arris by a rub or two on the stone. The cross piece or pieces on the top of the box are omitted for the sake of clearness.

Ornamental Arches

are those that have moulded soffits; and in such as the semi and segment, and in fact all that have the courses to one bevel, the moulding may be worked square, and applying the face mould cut in every respect similar to an arch with a square soffit. In this case one bed (the bottom one) will be square with the soffit, and the other very much wedge-shaped. The courses must be cut rights and lefts, but the key and two springing bricks must be wedge-shaped from both beds, otherwise they will want bedding up with large joints to fit the centre, and thus spoil the appearance of the arch.

When a camber, or any arch whose courses have different bevels, has to be moulded on the soffit, the bricks must *first* be bevelled and afterwards moulded, and, lastly cut to the required shape and length by the application of the face mould, as before described.

The Oriel Window

belongs peculiarly to ornamental brickwork (stone constructions being entirely excluded from this work), and we may add red brickwork. The first thing to be considered in connection with the oriel is its counterbalance. In all heavy projections in brickwork York flagging stones are employed; they are built into the main wall from which the projection starts, projecting to a distance suitable for the work. The weight of the projection on the stones is counterbalanced by the greater weight of brickwork on the other ends of the York slabs. But in the present case a girder or rolled iron joist, running in the direction of the wall line, and entering some 12 inches into the brick wall forming the side jambs, would have to be placed sufficiently low to allow the floor boards to pass over it. The flags and the weight upon them would be counterbalanced by the girder. The principle of counterbalance is known to bricklayers by the name of "tailing down."

The whole of the oriel (Fig. 72) as shown

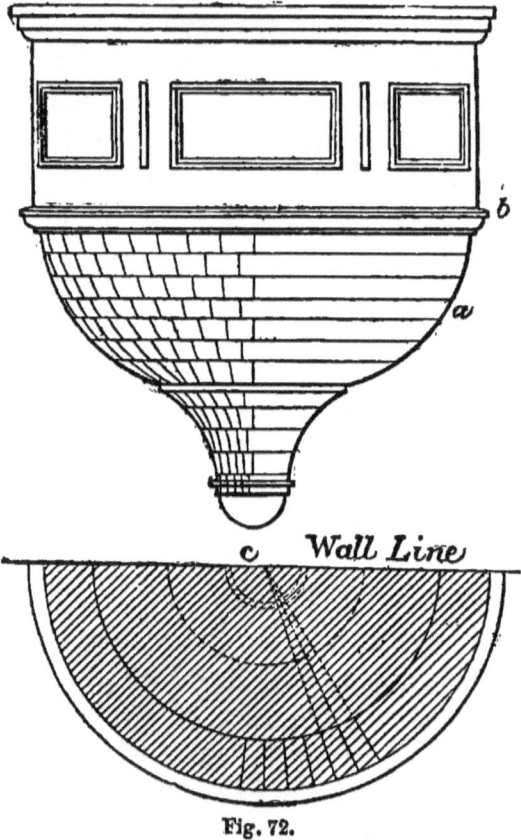

Fig. 72.

would be in brickwork, "gauged" and set in putty. The projecting courses, as the moulded string *b*, and the window-sill would be covered with 5-lb. lead, slightly projecting to form a drip for the water or rain.

The base here shown would be surmounted with mullions in brick or wood (most likely wood on account of its comparative lightness), and finished either with a semi-coned tiled roof or a balustrade. Windows of this type may be seen at Carlyle House, Chelsea Embankment; and the Agnew Picture Gallery, New Bond Street.

The bricklayer when setting out the work must strike all the successive courses from one point, c, regulating the length of the radius-rod for each course. Each course must radiate from c, as shown in plan, and the face of each brick be worked to the required sweep or curve. The bevels (which will be different for each and every course) will be obtained by placing the stock of the bevel on the line representing the bed, and bringing the blade to coincide with that portion of the curve representing the course we are about to cut. Let the bevel of the course marked a be required. Place the stock of the bevel on the third line below the moulded string b, and shift the blade until it fit the curve of the course a. The bevels for each course must be obtained in the same way. The plan in this figure may be considered as a horizontal section just above the string course b.

ORNAMENTAL GABLE OR PEDIMENT.

Figs. 73 and 74 are part front and end elevations of an ornamental gable or pediment. The moulding is composed of the members known as the ovolo, the cavetto, and the ogee. In ornamental brick copings it is usual to form the top fillet with two courses of red tiles, well soaked and closely and neatly set in cement, with the

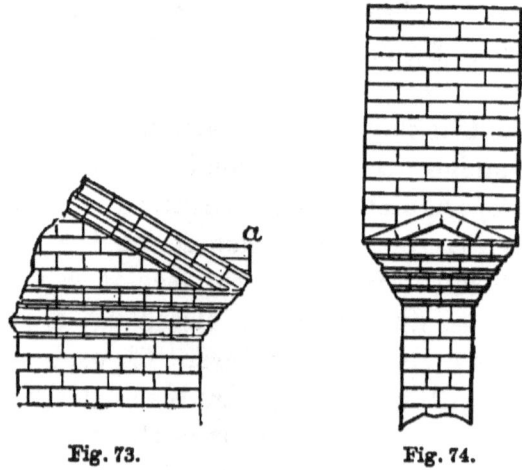

Fig. 73.   Fig. 74.

joints properly broken, as here shown. Sometimes lead is substituted for tiles. Here we have shown a gablet, *a*, but in practice the tiles are more frequently brought down to the bottom of the coping, the gablet being dispensed with.

Gothic Window.

Fig. 75 is a two-light ornamental Gothic window with 2-inch beaded or chamfered reveals. The whole of the work under the large arch would be recessed back from the general wall line. The side piers A and B for uniformity sake might be built in half bond, similar to that of the 9-inch mullion; but the proper bond would be to start from the reveal with a header and closer, the same as that shown on the reveal under the large arch. The tympanum is filled in

Fig. 75.

with 4½-inch work in 9-inch blocks, each block being made up of three bricks, and called "blocking courses."

The label or dripstone, *c e*, enclosing the large arch, for the sake of contrast might be in Portland stone. The whole of the work here shown, excepting the reveals of the large opening, might be in "gauged" work or in "axed" work; or the arches alone might be "gauged" or axed, with the tympanum filled in with good building bricks, selected for colour and shape and neatly pointed, making a very effective as well as economical ornamental feature.

The saddle-back springer on the mullion might with advantage be in stone. Windows of this kind may be built for cased frames with sliding sashes, but they are more generally built in neat work inside and out, with 9-inch jambs, grooved to receive lead lights. Ornamental brickwork is a subject in itself, that to adequately describe would require more space than can be given to it in a treatise of this dimension.

Mr. Ruskin, advocating its use, says: "Here let me pause for a moment to note what one should have thought was well enough known in England, yet I could not, perhaps, touch upon anything less considered—the real use of brick. Our fields of good clay were never given us to be made into oblong morsels of one size. They were given us that we might play with them, and that men who could not handle a chisel might knead out some expression of human thought. In the ancient architecture of the clay districts of Italy, every possible adaptation of the material is found, exemplified from the coarsest and most brittle kinds, used in the mass of the structure, to bricks for arches and plinths, cast in the most perfect curves, and of almost every size, strength and hardness; and moulded bricks wrought into flower work and tracery as fine as raised patterns upon china. And just as many of the finest works of the Italian sculptors were executed in porcelain, many of the best thoughts of their architects were expressed in bricks, or in the softer material of terra-cotta; and if this were so in Italy where there is not one city from whose towers we may not descry the blue outline of the Alps or Appennines—everlasting quarries of granite and marble —how much more ought it to be so among the fields of England."—*Stones of Venice,* vol. ii., p. 260.

Judging by the remarks in the above quotation, one is led to think that the brickmakers of mediæval Italy were more skilled in their craft, or at least happier in results, than their fraternity of modern times; for, with few exceptions, we have found moulded work wanting in that truthfulness of form which distinguishes cut or gauged work. Doubtless this, in great measure, is due to the large amount of unskilled and juvenile labour employed in our brickworks, to the careless manipulation of the work, and the hurried demand for the material. To be assured that true form *can* be obtained in ceramic wares, one has only to look at the Natural History Museum, London.

SECTION V.

ROOF-TILING, POINTING, Etc.

TILING.

TILING is a branch of the bricklayer's trade, and owing to the rage for red-brick buildings is now very much in use. One advantage of the tiled roof is that it is cool in summer and warm in winter, but on acount of their weight stronger timbers are required than for slates. The Broseley tiles are considered the best; they are $10\frac{1}{2}$ inches long, 6 inches wide, and $\frac{3}{8}$ of an inch thick, and have three nibs or projections at the head for hanging. Good tiles are fairly smooth and slightly vitrified. Those of a bright red or clayey colour, with no vitrification, are absorbent, and not so capable of resisting the weather. Six kinds are used in good work, viz. under-eaves or three-quarter tiles, plain tiles, hips and valleys, ridge tiles and tile-and-a half, the last being used for cutting up to valleys and hips, and forming gables, so as to do away with the half tile that would be required to break joint. Valley and hip tiles are purposely made to suit the angles of the roof. As the tiles come to the hand of the tiler he should throw out the straight ones to be used by themselves, while those that have a hollow bed should be also kept by themselves, as the straights will not lie close on the hollows. Good tiling is characterised by the tails of each course fitting closely upon the backs of the tiles in the course below them; by the cross joints or "perpends" running in straight and regular lines from eaves to ridge, the vertical joint between each two tiles coming immediately in the middle of the tile below them; by the hips and valleys being in the same plane as the sides of the roof of which they form a part. It is a common sight to see hips standing up above the roof, so as to have more the appearance of ridges than hips. As the tiles are ordered before the roof is on, the angles should be set out and sent to the tile-maker to insure getting them to the required angle. The contained angle of hip tiles is made 10° greater than the contained angle formed by the intersection at the hip of the two sides or planes of the roof, to allow for the tilt and the thickness of the two eaves-tiles. For the same reason the valley-tile is made 10° more than the re-entering angle of the roof. In our experience we have frequently found that the contained angle has been guessed at or obtained by some "rule of thumb," and with the consequence that generally ensues from such work, viz. that the angle contained within the hip tile has been either too acute or too obtuse.

Tiles are either laid dry on close boards, with battens above for hanging them, or on open battens, in which case they should be bedded in lime and hair mortar. The most modern and improved way of hanging is shown in Fig. 76. The boards are 6 inches wide and are feather-edged, the top edge being $\frac{3}{4}$ of an inch thick. Here we have a boarded roof without battens, and one that will keep out the weather if the tiles should get broken, for the rain would cause the wood to expand, and thus tighten the joints of the boards, to the exclusion of all rain. The first course—the eaves and under-eaves—should be bedded in hair mortar. The "lap" (the distance that the tail of the third tile overlaps the head of the first) should be

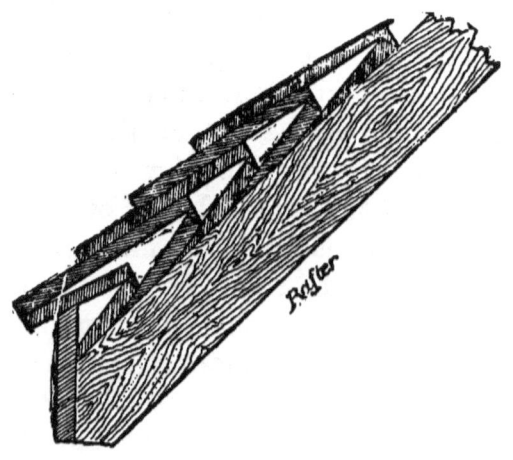

Fig. 76.

3 inches. The "gauge" (the distance between the tails of two consecutive courses) can always be obtained by dividing the length of the tile (measured from the under side of the hanging nibs) less the lap by two. Thus, $(10\frac{1}{2} - 3) \div 2 = 3\frac{3}{4}$, the "gauge."

Roofs having Different Pitches.

When roofs of different pitches intersecting in hips and valleys occur, the tiler has generally a

ROOF-TILING, POINTING, ETC. 95

deal of trouble, and consequent waste of time, through carpenters frequently insisting upon intersecting the battens; and very often after much time has been wasted, and a portion of the tiling done, it is found necessary to tear off all the battens to correct the error.

The following rule will prevent such an error. Draw the plan of the two roofs (Fig 77), of

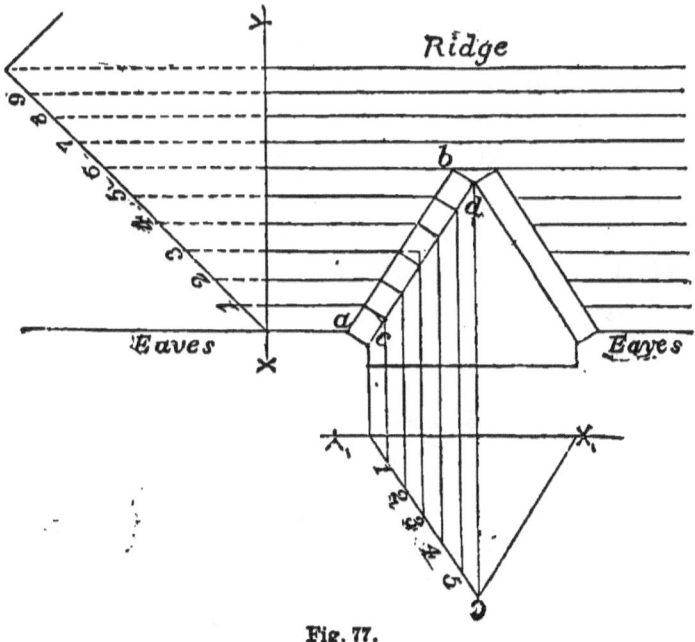

Fig. 77.

different pitch, and from the centre of the valley set out two parallel lines, *a b, c d*, representing the true width of the tails of the valley tiles, which is from 1½ to 2 inches. On *x y* at right angles with the eaves of the main roof draw its section, on which set out the gauge 1, 2, 3, &c.,

and drop lines square with xy and intersecting the line ab. From these points of intersection square the short lines across the valley, and from where they intersect the parallel cd draw lines square with $x'y'$ and intersecting a section of the smaller roof. The distance between any two points on y' G will be the "gauge" for the smaller roof. The line 3 on each section is drawn to their intersection, which is *not* in the centre of the valley, but very much on one side of it, thus proving the *popular* error of intersecting the battens in the middle of the valley.

The "gauge" for hips should be obtained in the same way, excepting that the parallel lines, ab, cd, must be the same distance apart as the extreme points of the tail of the hip tile, measured in a straight line from point to point square with the hip.

To obtain the Necessary Angle of Hip of Valley Tiles.

Draw ab, Fig. 78, the plan of the hip, and erect a perpendicular, ac, the true height of the top of the hip. Draw a line from c to b, and the angle abc will be the true inclination of the hip. Draw ed square with ab, cutting the eaves, and from f a line square with cb; with this as radius, from the point f draw the semicircle, and from where it cuts ab draw the lines eg, dg; egd is the angle required for the hip tiles, or in other words it is a section or cut through the roof at right angles with the hip. The angle for valley tiles is obtained in the same way, remem-

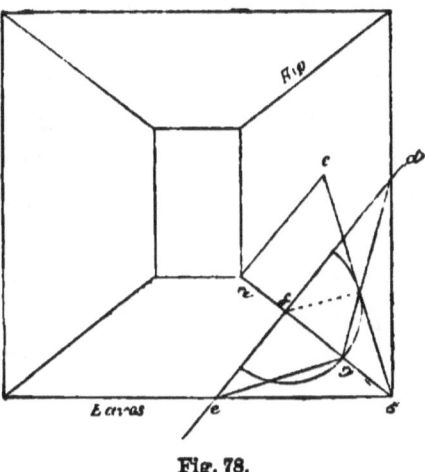

Fig. 78.

bering that the hip is a salient angle and the valley a re-entering angle.

Pointing.

Pointing is divided into two classes, tuck-pointing and flat-joint pointing. In tuck-pointing the joints of the brickwork are filled in with mortar or stopping, of generally the same colour as the bricks, and rubbed down to a level surface with a piece of sacking or soft brick of the same colour as the work, and a putty joint made of lime and silver-sand placed upon it. Stone lime should be used for outside work.

The mode of working is to have a parallel rule from 8 to 10 feet long, 5 inches wide, and $\frac{1}{2}$ an inch thick, with one feather edge and four cleats $\frac{3}{16}$ of an inch thick tacked on to the back to afford room for the putty that is cut off to fall through. The putty is spread out on the rule from which the bricklayers, one at each end, take it off with their jointers, and with the rule against the wall, working on the top edge, transfer it to the wall. The ragged edges are then cut off with the Frenchman or knife, and the loose particles brushed off with a soft brush. Tuck-pointing is not suitable for outside work, as the putty joints projecting beyond the general surface arrest the weather and are consequently soon destroyed, unless protected by heavy projections.

Flat-joint Pointing.

This is the most general and durable kind of pointing. It should be made up of washed sand and stone lime several days at least before using it, that it may by the porcess of retempering acquire toughness, which will add very much to its durability and facility of working. The joints should be finished flush with the work (excepting in "weather-jointing," when the top of the joint should be kept back $\frac{3}{8}$ of an inch, and the bottom flush to shed the rain) and neatly cut off top and bottom with the Frenchman, and brushed off. To ensure good pointing, the work should be well raked out and wetted not sparingly. If the joints are deep they should be filled in by going over them twice with tolerably stiff mortar to prevent cracking, and the work done with *pointing trowels*. Jointers should not be used under any pretext. In first-class work the pointing is done as the work proceeds during erection, and forming one body with the building will, if the mortar be good, last for many years.

Malm work for tuck-pointing is generally stopped in with mortar, coloured with yellow ochre (2lbs. of ochre to each hod of mortar), but it will be found best to use *no* colour in the stopping, as by its earthy nature it very much injures the setting and hardening properties of the lime, which in a great measure accounts for so much pointing perishing soon after it is done. Stop the work in with good mortar, as described in flat-joint pointing, and rub it down with a soft malm, leaving the dust on the work, and with a soft stock brush go over it lightly with hot alum water. One pound of alum to 3 gallons of water.

White Suffolk bricks for tuck-pointing, are treated in the same way, rubbing the work with a soft white Suffolk instead of with a malm.

Red work for tuck-pointing is stopped in with mortar coloured with Venetian red and Spanish brown, with sometimes a little vegetable black added. The colour of the stopping must be determined by the colour of the bricks, so as to match them. It is best to avoid colouring the bricks, but when stopped in rub them down with a soft brick, and apply alum water or white copperas, as already described. One pound of copperas to 3 gallons of water. The appearance of red brickwork is often spoilt through the application of colour.

To clean down red work, mix a pint of spirits of salts with a pailful of water. This applied with a stock brush will leave the work clear of all lime spots, &c. It may be done on work recently erected, in which the joints have been struck during erection, and without injuring them.

Copperas is very much used in connection with stock work, especially when the bricks are inferior or of a bad colour. One pound of green copperas is melted down with every 5 gallons of water. It should be mixed several days before required, and enough made to finish the job, that it may be all one colour. A small nob of lime mixed with the copperas very much heightens its colour. The copperas should be tried on the work to match it before being generally used, and weakened down by the addition of water if found necessary.

Burning Clay into Ballast.

The use of burnt ballast is increasing every day, both for purposes of mortar and concrete. The chief reason for this is its cheapness in comparison with the cost of sand, for while sand costs from 5s. to 7s. a cube yard, varying according to the locality, burnt ballast can be produced, including all materials and digging of clay, with a run of about 60 yards, at 2s. 6d. a cube yard. While we *reiterate* that for mortar nothing better than clean sand of a sharp angular grit can be used, we do not wish to be understood as condemning the use of burnt ballast. Thoroughly burnt and cool, with the large aggregations (sponge-like lumps whose parts touch each other here and there, and are held in contact by vitreous matter) broken up, and the whole mixed with a fair proportion of Thames ballast or clean gravel (see previous remarks on this subject in Article on Concrete), is capable of making a good concrete, for the absorbent nature of the ballast attracting the silicates of the cement or lime, which entering the pores form so many threads or ties binding the whole mass together, and unlike Thames ballast, with its non-absorbent and smoothly water-worn surfaces, which simply beds itself in the matrix with comparatively little adhesion.

Stiff or strong clay, just as it is dug up, is the best for burning, as it requires the least firing and will make the best ballast. The heap is commenced by forming a cone of clay, about 3 feet in diameter and 5 feet in height, formed round a piece of pole placed on end as a centre. Fires are then made round the cone by placing bricks on edge forming a channel leading up to the centre. These are filled with wood and coal, and covered over and cased with a layer of clay about 6 inches thick before lighting. As the fire burns through it must be drawn down, which is done by means of long-handled prongs made specially for the work, and strewn with small coal called " slack," and covered with another layer of clay. The thickness of the layers of clay may be increased as the work proceeds, until they become from 18 to 24 inches, not forgetting the sprinkling of "slack" on each layer of clay. Care must be taken that the fire be drawn down, as it naturally draws to the top, and the unburnt portions thrown up into the fire. When the clay is thoroughly burnt the fire will go out.

BUILDING ADDITIONS TO OLD WORK.

When building additions to old buildings, it frequently occurs that the old work is found to be considerably out of perpendicular, generally overhanging. In such a case it is best to carry up with the new work, just where it joins with the old, a pier or pilaster, forming a break in the wall line, which will enable the bricklayer to keep the new work upright and hide the fault of the old, which otherwise would be exposed by junction with the new. The projection of the pilaster will of course be regulated by the amount that the work is out of the upright.

FIRE-PROOF FLOORS.

Fire-proof floors are now very rarely constructed in bricks, being almost entirely superseded by tile arches brought to a level with concrete, or constructed with rolled joists and concrete alone, or with cement and breeze, but more generally with Dennett's Patent, which is a concrete composed of broken bricks and gypsum. But in very large warehouses, and where great weights have to be carried, the fire-proof floors are still constructed with brick rings carried on rolled girders.

SECTION VI.

APPLIED GEOMETRY.

Geometry of all studies is to the artisan the most attractive and useful. The problems given here are such as may be applied by the bricklayer to every-day practice, and therefore come within the meaning of the term *applied geometry*. But we would advise the young artisan not to rest satisfied with a knowledge of the few problems given herein, but to take up the subject as a separate study, and familiarise his mind with its principles, so as to be able to apply them generally and with understanding.

To draw a square whose superficial area shall equal the sum of two squares whose sides are given.

Let A B (Fig. 79) be the given sides. Draw the lines C D, E F at right angles, and from G set off G H equal to A, and G K equal to B: a line drawn from H to K will be the side of the required square. On G K complete the square G M, N K; and on G H the square H L E G; and on H K the square H K O P. The area of this square will equal the combined areas of the two smaller

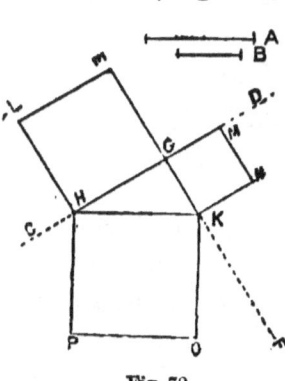

Fig. 79.

squares. To make this more clear, suppose the line A to be 8 inches and B 6 inches. The square of A would be 8 × 8 equal to 64; and the square of 6 would be 6 × 6 equal to 36, which added to 64 makes 100. By drawing A and B square with each other and joining their extremes with a straight line, we will find that line to measure exactly 10 inches, and the square of that will be 100.

The principle of this problem is that a square erected on the hypothenuse (the longest side) of a right-angled triangle is equal to the sum of two squares, erected on the base and perpendicular of the same triangle. Its application to practice is shown in the article on "Setting out Building."

To draw a right-angled triangle, base 1½ inches, height ½ inch.

Draw a semicircle of 1½ inch diameter (Fig. 80), and from d erect the perpendicular $d\ e$: a line drawn from e, ½ inch above the base line $a\ c$, will cut the semicircle in b; lines drawn from a and c to b will form

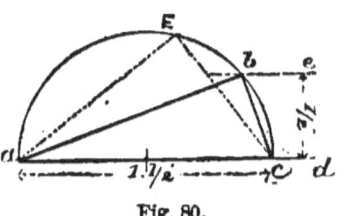

Fig. 80.

the required triangle. The principle of this is that all triangles within a semicircle are right-angled triangles. If the lines be drawn from $a\ c$ to E or to any other point in the semicircle, we shall get a right-angled triangle. Its practical application is seen in the article on "Setting out Building."

To draw an arc by cross-sectional lines.

On *a b*, the span (Fig 81), erect the perpendiculars, *d e*, equal to twice the required rise. Divide *a e* into any number of equal parts, 1, 2, 3, 4, and *e b* into the same number of parts, and draw cross-sectional lines as shown. A curve traced through the intersections will be the required arc.

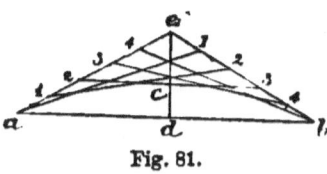

Fig. 81.

Another method practised (we do not recommend its use) sometimes by carpenters for getting out turning-pieces for the bricklayer. Span 6 feet, rise $1\frac{1}{2}$ inch. Divide the span into a number of equal parts, say six, and from the points erect perpendiculars, measuring upward $\frac{1}{2}$ inch on the first, an inch on the second, and $1\frac{1}{2}$ inch on the third, which in this case is the centre line. Treat the other half of the span in the same way, and with a flexible straight-edge fixed at the springing points *a b* (Fig. 81) force it upward until it stand over the distance marks on the perpendiculars, and with a pencil trace the arc or curve.

The foregoing methods do away with the necessity of laying down a large platform and getting out a long radius-rod; the camber, for instance, which is the segment of a circle described by a radius-rod of 70 feet $2\frac{3}{8}$ inches in length.

To describe a flat arc (camber for instance) by mechanical means.

Let $a\,b$ (Fig. 82) be the cord of the arc. Bisect

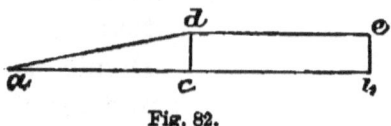

Fig. 82.

$a\,b$ at c by the perpendicular $c\,d$, and make $c\,d$ equal to the height of the segment. Draw $d\,e$ parallel to $a\,b$, and make $d\,e$ a little larger than $a\,d$. This template should be got out of a piece of timber, and by moving the whole of the template, so that the two edges $d\,a$ and $d\,e$ may slide on two pins, a and d, the angular point d of the template will describe the segment required, and if the pin be taken out of a and put in the point b, the other portion $d\,b$ of the segment $a\,d\,b$ will be described in the same manner. This method should be practised in preference to the methods previously described.

To find the joints of a flat arch without using the centre of the circle of which the arc is a part.

Having determined the number of voussoirs or

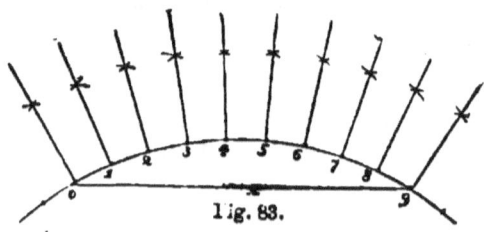

Fig. 83.

"courses," 1, 2, 3, 4, &c. (Fig. 83), from these points

erect perpendiculars by intersecting arcs; these perpendiculars represent the joints. We need hardly to say that the practical application of this problem is to enable the workman to draw the courses or voussoirs in an arch similar to that given in the previous problem.

To draw the joints of a semi ellipse arch with mathematical accuracy.

The point D (Fig. 84) is the middle of the arch,

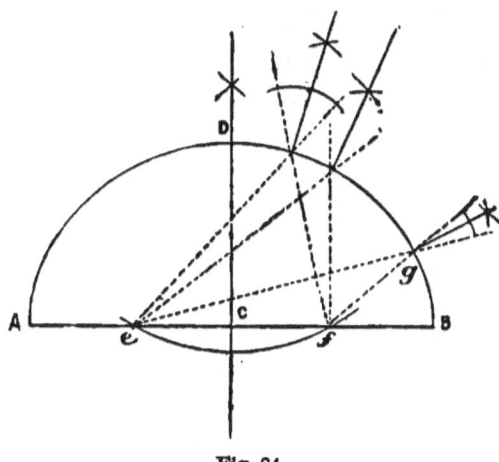

Fig. 84.

and the point c the middle of the springing line. With the distance C A or C B, from the point D describe an arc cutting A B at *e*, and also at *f*; *ef* are the foci. Let a joint be required at *g*. From *e* and *f* draw lines passing through *g*, and bisect the angle they make with each other, and from the point *g* erect a perpendicular, which will represent the required joint. The other joints are obtained in a similar manner.

To find the invisible arch contained in a camber.

Bisect the springing line *a b* (Fig. 85) with the perpendicular *c d*, and produce the skewback *h b* until it cut the perpendicular in *c*. From *c*, with distance *c b* draw the arc *a d b*, and with distance *c g* its concentric arc *g f h*. *a g h b* is the invisible arch. The soffit of the camber below the arc *a d b* is upheld by the cohesion of its parts with the invisible arch. Here we would add that bricklayers have no fixed rule to determine the angle of skewback for the camber, some giving 4½ inches skewback for all openings, others 6½ inches, and in many cases giving a skew of from ¾ to 1 inch for every foot that the opening is wide; as 3 inches for 3 feet, 4 inches for 4 feet, and so on. We would advise that the skew or angle of thrust should never exceed 6 inches, for as the skew becomes more acute the carrying strength of the camber becomes less, in consequence of the invisible arch contained therein being thrown higher up, as shown by the middle arc struck from *k* with distance *k b*.

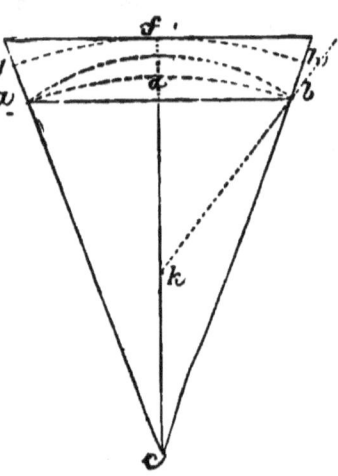

Fig. 85.

APPLIED GEOMETRY. 109

Any two straight lines given to determine a curve by which they shall be connected.

Let *a b, c d* (Fig. 86), be the given lines, and *c b*

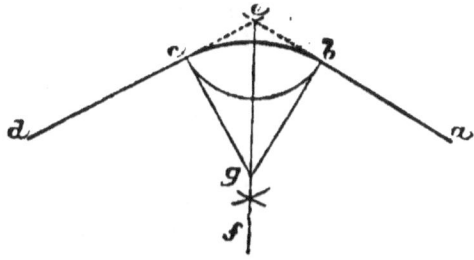

Fig. 86.

the points to be connected. Produce the lines until they meet in *e*; bisect the angle *c e b* with the line *e f*; from *c* and *b* draw lines at right angles to *a b* and *c d* meeting *e f* in *g*. From *g*, with distance *g c* or *g b* describe the connecting curve. The given lines may be taken as two brick walls that have to be connected or formed with a round corner.

Fig. 87 is an example in which the given lines

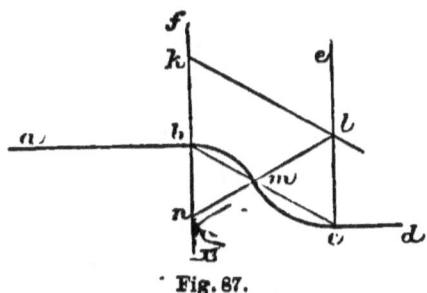

Fig. 87.

are parallel. From point *b* draw *f x* at right angles with *a b*; and from *c, c e*, at right angles with *c d*. On *f* mark a point *k* any distance from *b* less than B C. Draw *k l* through *k* parallel to *b c* and cutting *c e* in *l*. From *l* as centre with the distance *l c*, which is equal to *b k*, describe the arc *c m*. Join *l m* and produce it in the same straight line towards *m* to meet *f x* in *n*. From *n* as centre, with the distance *n b* or *n m*, describe the arc *b m*. The given straight lines *a b*, *c d* are connected by the curve *b m c*.

If the given straight lines are not parallel, but would meet if one or both were produced, as *g h* (Fig. 88), produced meets *a b* in *a*, forming the

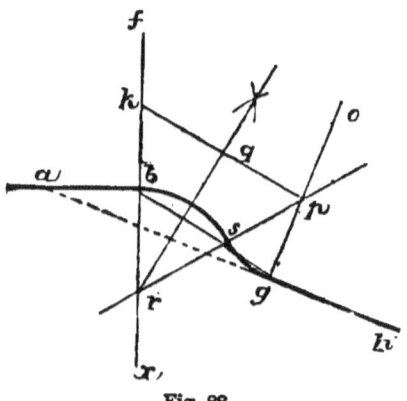

Fig. 88.

small angle *g a b*, draw, as before, *f x* and *g o* at right angles to *a b* and *g h* respectively. Take any point, *k*, in *b f*; make *g p* equal to *b k*, and join *k p*. Bisect *k p* in *q*, and draw *q r* perpendicular to *k p*, meeting *f x* in *r*. Join *r p*, and from *p* as centre, at the distance *g p*, describe the arc *g s*, meeting *r p* in *s*. Then from the centre *r*, at the distance *r b* or *r s*, describe the arc completing the

APPLIED GEOMETRY. 111

curve *b s g*, by which the given straight lines *a b*, *g h* are connected.

To find the form or curvature of a raking moulding that shall unite correctly with a level one.

Let *a b c d* (Fig. 89) be part of the level

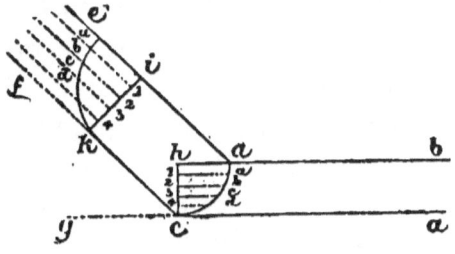

Fig. 89.

moulding (which we will here suppose to be an ovolo or quarter round), *a* and *c* the points where the raking moulding takes its rise on the angle, *f c g* the angle the raking moulding makes with the level one. Draw *c f* at the given angle, and from *a* draw *a e* parallel to it; continue *b a* to *h*, and from *c* make *c h* perpendicular to A *h*. Divide *c h* into any number of equal parts, as 1, 2, 3, 4, and draw lines parallel to *h* A, as $1^a, 2^b, 3^c, 4^d$; and then in any part of the raking moulding, as *i*, draw *i k*, perpendicular to *e a*, and divide it into the same number of equal parts as *h c* is divided into; and draw $1^a, 2^b, 3^c, 4^d$, parallel to *e a*. Then transfer the distances $1^a, 2^b, 3^c, 4^d$, and a curve drawn through these points will be the form of the curve required for the raking moulding.

The method here shown is for an ovolo, but it would be just the same for any other formed moulding, as a cavetto, ogee, &c. This problem can be applied in the construction of pediments in " gauged " work.

To describe an ellipse by means of a carpenter's square and a piece of notched lath.

Having drawn two lines to represent the diameters of the ellipse required, fasten the square so that the internal angle, or meeting of the blade and stock shall be at the centre of the ellipse. Then take a piece of wood, or a lath, and cut it to the length of half the longest diameter, and from one end cut out a piece equal to half the shortest diameter, and there will then be a piece remaining at one end equal to the difference of the half of the two diameters. Place this projecting piece of the lath in such a manner that it may rest against the square on the edge which corresponds to the two diameters; and then turning it round horizontally, the two ends of the projection will slide along the two internal edges of the square, and if a pencil be fixed at the other end of the lath it will describe one quarter of an ellipse. The square must then be moved for the successive quarters of the ellipse, and the whole figure will thus be easily formed. This method is on the principle of the trammel. There are several other ways of drawing an ellipse, but for these the reader must be referred to a work on geometry.

APPLIED GEOMETRY. 113

*To draw a Gothic of any given height and span;
or, in other words, an Ellipse Gothic.*

Let A B (Fig. 90) be the span and C D the height. Draw the line A B and bisect or centre it at *c*;

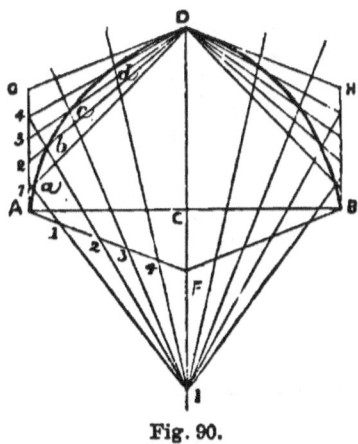

Fig. 90.

draw C D, and make C *i* equal to C D. Divide C D into three equal parts, and draw A G, B H parallel with C D, and equal to two-thirds ($\frac{2}{3}$) of C D. Make C F equal to one-third of C D, and draw A F, F B. Divide A F into any number of equal parts, 1, 2, 3, 4, and from *i* draw *i*1, *i*2, *i*3, *i*4. Divide A G into the same number of parts as A F, and draw 1D, 2D, 3D, 4D, and the intersection of lines will give the points in the curve, which must be drawn by hand. The other half must be found in the same way.

To draw the arch bricks of a Gothic arch, that is for the curve in the previous problem.

Having formed the angles C D G and C D H as before, from D (Fig. 91) draw D L perpendicular

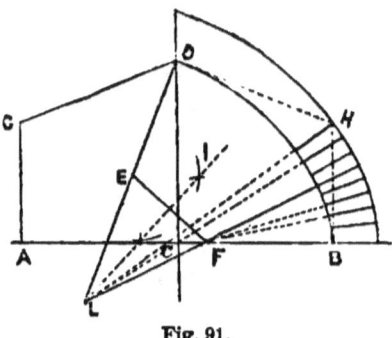

Fig. 91.

to D H. Make B F and E D each equal to B H; join E F, and from the middle of E F draw *i* L perpendicular with E F. Draw L F, L and F are the points from which the joints of the arch will radiate.

To find the radius of any arc or arch, the rise and span being given.

Let *a b* represent the span, *c d* the rise; *a b* equal 4 feet, *c d* 2 feet. *a c* (half the span) multiplied by itself will be 2 × 2, or 4 feet; divided by *c d* will be $\frac{4}{2}$, or 2 feet. *c d* added to this will be 4 feet, which divided by 2 will give 2 feet as the length of radius that will describe the required arc whose span and rise are given. In this case we have chosen a semicircle for the sake of simplicity and self-demonstration, but the rule may be applied to any arc of any circle. In mathematical formula our calculation would stand thus:

$$\left(\frac{a\,c^2}{c\,d} + c\,d\right) \div 2 =$$ the length of radius required. Or in plain words $a\,c$ square, divided by $c\,d$, plus $c\,d$ divided by 2 equal the length of radius. In the above explanation we have gone out of the beaten track for the purpose of making the rule clear to those of our readers who may not be familiar with trigonometrical and algebraic expressions.

It will, however, be recognised by some as the square of half the cord divided by the versed sine, plus the versed sine divided by 2 equal the radius.

For mensuration of brickwork the Author refers the reader to Mr. Hammond's "Practical Bricklaying," forming vol. 189 in this series.

INDEX.

Additions to Old Work, 102.
Angle of hip or valley tiles, to obtain, 96.
Angle of skew, 50.
   of strain, 29.
Apertures, 26.
Arc, to draw by cross sectional lines, 105.
Arches, 46.
   cutting of, 64.
   principles of, 47.
   whose courses have different bevels, 85.
Axed work, 63.

Base, Bond of, 27.
   treatment of, 27.
Battering jamb, 38.
Bats, 30, 34.
Bedding board, 66.
Bends, 12.
Berkshire builders, 17.
Birdsmouthed key, 71.
   objection to, 71.
Blocking courses, 89.
Blue lias, 9.
Boaster, 65.
Bond of brickwork, 20.
   underrated, 20.
   of footings and walls, 22.
   Gwilt on, 21.
   Smeaton on, 20.
Boyd's flue-plates, 45.
Bricks, 16.
   characteristics of good, 19.
   differences in sizes of, 39.
   case hardening of, 62.
   wetting of, 28.

Brick groins, 58.
Brickwork, 2.
   characteristics of good, 40.
   good, samples of, 16.
Broken bond, 27.
   cause of, 27.
   in Flemish, 33.
Brondesbury bridge, 55.
Broseley tiles, 92.
Building new work into old, 40.
Bull's-eye, 65.
Burning clay into ballast, 100.
Burnt ballast, 9.
Buttering joints, evil of, 34.

Camber arch, 64, 68.
   invisible arch in, 108.
   mould, 69.
   to describe by mechanical means, 69.
   to take off lengths and bevels of courses, 68.
Carved work, 64.
   gauged-work, composition for setting, 60.
Catenary curve, 48.
Cement-testing machine, 10.
Centre for niche head, 81.
Ceramic wares, 91.
Chalk lime, 14.
Chimney bond, 34.
Chimney stacks, 34.
   walls of 4½ inches, 34.
Closer or Closure, 22, 31.
Colour in stopping, objection to, 99.
Concrete, 2.
   for filling in terra-cotta, 8.

INDEX.

Concrete, Mr. Reid on, 8.
    "packing," 7.
    proportion of ingredients, 8.
    quantity of water in mixing, 8.
    specification of mixing, 6.
    thickness of, 5.
Construction of arches, 55.
Copperas, 99.
Coring holes, 43.
Counterbalance, 85.
Coursing joints, 50.
    mould, 52.
Cross joints, 24.
Cutting-shed, 64.

DENNETT'S PATENT, 102.
    Development of soffit of skew arch, 51.
Dipping, 63.
Dips in drains, 12.
Dip-trap, 13.
Doors, positions of, 26.
Doulton's terra-cotta flue pipes, 45.
Drains, laying of, 11.
    fall of, 11.
    cause of stoppage, 12.
    ventilation of, 12.
Drain-pipes, sizes of, 13.
Drawing and cutting arches, 64.
Dutch bond, 37.
    advantages claimed for, 38.

EARTHENWARE TRAPS, 13.
    Ellipse, to describe, 112.
Ellipse Gothic arch, 72.
    to describe by cross sectional lines, 113.
Enamelled bricks, 18.
English garden wall bond, 35.
English bond in chimney stacks, 34.
Excavations, 2.
Extrados, 66.

FACE MOULD, 65, 66.
    Fan-groining, sample of, 60.
Fareham bricks, 17, 61.
Fire bricks, 18.

Fireclay, 19.
Fireplace for register stove, 44.
Fireproof floors, 102.
Flat joint pointing, 98.
Flashing to chimney-stacks, 35.
Flat arch, to find joints of, 106.
Flemish, 22, 31.
Flemish garden wall bond, 35.
Flues, 41.
    building of, 43.
    down draught, 42.
    disadvantage of too large sectional area, 42.
Flues, sizes of, 44.
Flushing, 28.
Footings, 6, 22.
Forcing-rods, 12.
Formation of centre for niche head, 81.
Foundations, 1.
Freestone lintels, cause of fracturing, 48.

GABLET, 88.
    Gauge of tiles, 95, 96.
    Gauge-rod, 25.
Gauged work, 17, 61.
Gault bricks, 17.
Geometry, 103.
Gothic arches, 63, 70, 71.
    to draw arch bricks of, 114
    on circle arch, 77.
    vaulting, 58, 59, 60.
    window, 89.
Grizzles, 16.
Groined vaulting, 58.
Ground blue lias, 2.
Ground-damp, 2.
Grouting, 28, 40.

HEADING BOND, 28.
    Herring-bone bond, 36.
Hip tiles, 92, 93, 96.

INSPECTION HOLES, 10.

JAMB, 26.
    Jointers, 98.

KING CLOSURE, 25.
   Kneeler, 88.

LARRYING UP, 41.
   Level or datum, 2.
Lime, 14.
Lines of force or thrust, 55.
Line of frontage, 3.
Line of radii, 55, 57.
Lines, to connect by means of a curve, 109.
Lintels, 48.
London clay, 6.
Long skew arches, treatment of, 51.
Lump lias, 14.

MADE-UP GROUND, 6.
   Malms, 17, 61.
   action of London smoke on, 62.
Man-hole, 12.
Mortar, 14, 28.
Moulded courses, 83.
   work, 91.
Mullions, 74, 89.

NICHE, 79.
   hood, 81.
   mould, 83.
   length and bevel of courses to, 82.

OLD ENGLISH BOND, 21.
Open soil-pipe, 12.
Operculum or channel-pipe, 12.
Oriel window, 85.
Ornamental arches, 84.
Ornamental brickwork, 79, 90.
Ornamental gable or pediment, 88.

PARGETTING, 34.
   Paving, 36, 37.
Perpends, 39.
Philological School, 15.
Picked stocks, 16.
Place bricks, 16.
Plain arches, 49.
Plan of skew arch, 50.

Plinth, 27.
Pointing, 39, 97.
Polychrome bricks, 15.
Portland cement, 2, 6.
Portland cement concrete, 7.
Pressed bricks, 20.
Principle of ordinary skew arch, 54.
Projecting courses, 86.
Purpose-made brick, 57.

QUEEN ANNE'S STYLE OF ARCHITECTURE, 79.
   revival of, 14.
Quoin, 4, 32.

RADIUS-ROD, LENGTH OF, FOR CAMBER, 105.
   to obtain by formula, 114.
Raking moulding, 111.
Red brickwork, 14.
   to clean down, 100.
Red building bricks, 17.
Relieving arches, 48.
Reveal, 26.
   bond of, 27.
Reversing the bond, 33.
Right-angled junctions, 12.
Roman tile, 30.
Roofs of different pitch, 94.
Rosendale cement, 57.
Ruabon clay, analysis, 20.
   bricks and terra-cotta, 20.
Rubbers, 17, 61.
Ruskin, influence of on red-brick designs, 15.
   advocacy of ornamental brickwork, 90.

SAND, 9, 14.
   Scheme, 73.
Scheme arch, 75.
"Scotch," 65.
Scriber, 65.
Section-box, 84.
   mould, 78.
Section of niche, 79.
Semi-ellipse arch, 73, 107.
Semi and segmental arches, 66.
Semi-Gothic arch, 76.

Setting, 63, 64.
Setting out and cutting, 62.
Setting out building, 2.
Setting out the bond, 26.
Sewer gas, 12.
Sewers, 57.
Sharp bends in flues, evil of, 42.
Shippers, 17.
Site, 1.
Skew arch, 49, 52.
Skewback of camber, 103.
Smoky flues, 41.
Snapped headers, 31, 34.
Soakers, 35.
Soffit-mould of Gothic on circle, 78.
Stacks in 4½-inch walls, 34.
Staffordshire blue bricks, 18, 52.
Stock bricks, 16.
Stockwork, 61.
Stone lime, 14, 63.
Stone strings, 28.
Stopping in pointing, 97, 99.
Stourbridge fire bricks, 19.
String courses, 30.
Stuccoed buildings, 15.
Subsoil, 5.
Surface concrete, 2.

Tailing down, 86.
Taking off bevels, 65.
Templates and strings, 30.
Testing cement, 10.
Teynham bricks, 17.
Thames ballast, 10, 101.
Thick and thin joints, 28.

Three-quarter stretcher, 24, 27.
Tiled roof, advantage of, 92.
Tiling, 92.
characteristics of good, 92.
Tile fillet, 88.
Tiles, characteristics of good, 92.
improved method of hanging, 93.
Timber foundation, 57.
T. L. B. Rubbers, 61.
To find the radius of any arc or arch, 114.
Tools for arch cutting, 64.
Toothings, 39.
Transverse joints, 23.
Transversing the courses, 65.
Triangle, 104.
Tuck-pointing, 97, 99.
Tumbling in buttresses, &c., 38.
Tympana of arches, 36, 89.

Valley tiles, 96.
Valleys, 94.
Various bonds, 34.
Venetian arch, 74.
Voussoirs, 47, 65.

Wall in Flemish face and English back, 32.
Washed stocks, 16.
Water conduit, 56.
Weather-jointing, 98.
White Suffolks, 18, 62.
Windows, 27.
Wing gatherings, 41, 44.
Withes, 34, 35, 45.

THE END.

PRINTED BY J. S. VIRTUE AND CO., LIMITED, CITY ROAD, LONDON.

Uniform with this volume, price 1s. 6d.

THE RUDIMENTS OF PRACTICAL BRICKLAYING.

In Six Sections:—General Principles of Bricklaying—Arch Drawing, Cutting, and Setting—Different Kinds of Pointing—Paving, Tiling, Materials—Slating and Plastering—Practical Geometry, Mensuration, &c. By ADAM HAMMOND. Illustrated with Sixty-eight Woodcuts. Fifth Edition, carefully Revised, with Additions.

"This is the work of a practical bricklayer, and is intended for the junior members of the important, if laborious, profession to which the author belongs. It is full of details concerning all the parts of the shell of a building, from foundation to tiles. To any workman anxious after improvement this volume will prove a valuable investment."—*Iron.*

"Contains a considerable amount of practical information, with sound instructions on general matters, and useful recipes connected with both brickwork and plastering."—*British Architect.*

"Mr. Hammond's practical treatise will be found of great value to students."—*Building News.*

"Any young bricklayer who reads Mr. Hammond's book carefully will become a proficient craftsman,"—*English Mechanic.*

CROSBY LOCKWOOD & CO.,
7, STATIONERS' HALL COURT, LONDON, E.C.

Weale's Rudimentary Series.

LONDON, 1862.
THE PRIZE MEDAL
Was awarded to the Publishers of
"WEALE'S SERIES."

A NEW LIST OF
WEALE'S SERIES
RUDIMENTARY SCIENTIFIC, EDUCATIONAL, AND CLASSICAL.

Comprising nearly Three Hundred and Fifty distinct works in almost every department of Science, Art, and Education, recommended to the notice of Engineers, Architects, Builders, Artisans, and Students generally, as well as to those interested in Workmen's Libraries, Literary and Scientific Institutions, Colleges, Schools, Science Classes, &c., &c.

☞ "WEALE'S SERIES includes Text-Books on almost every branch of Science and Industry, comprising such subjects as Agriculture, Architecture and Building, Civil Engineering, Fine Arts, Mechanics and Mechanical Engineering, Physical and Chemical Science, and many miscellaneous Treatises. The whole are constantly undergoing revision, and new editions, brought up to the latest discoveries in scientific research, are constantly issued. The prices at which they are sold are as low as their excellence is assured."—*American Literary Gazette.*

"Amongst the literature of technical education, WEALE'S SERIES has ever enjoyed a high reputation, and the additions being made by Messrs. CROSBY LOCKWOOD & SON render the series more complete, and bring the information upon the several subjects down to the present time."—*Mining Journal.*

"It is not too much to say that no books have ever proved more popular with, or more useful to, young engineers and others than the excellent treatises comprised in WEALE'S SERIES."—*Engineer.*

"The excellence of WEALE'S SERIES is now so well appreciated, that it would be wasting our space to enlarge upon their general usefulness and value."—*Builder.*

"The volumes of WEALE'S SERIES form one of the best collections of elementary technical books in any language."—*Architect.*

"WEALE'S SERIES has become a standard as well as an unrivalled collection of treatises in all branches of art and science."—*Public Opinion.*

PHILADELPHIA, 1876.
THE PRIZE MEDAL
Was awarded to the Publishers for
Books: Rudimentary, Scientific,
"WEALE'S SERIES," ETC.

CROSBY LOCKWOOD & SON,
7, STATIONERS' HALL COURT, LUDGATE HILL, LONDON, E.C.

WEALE'S RUDIMENTARY SCIENTIFIC SERIES.

** The volumes of this Series are freely Illustrated with Woodcuts, or otherwise, where requisite. Throughout the following List it must be understood that the books are bound in limp cloth, unless otherwise stated; *but the volumes marked with a ‡ may also be had strongly bound in cloth boards for 6d. extra.*

N.B.—*In ordering from this List it is recommended, as a means of facilitating business and obviating error, to quote the numbers affixed to the volumes, as well as the titles and prices.*

CIVIL ENGINEERING, SURVEYING, ETC.

No.
31. *WELLS AND WELL-SINKING.* By JOHN GEO. SWINDELL, A.R.I.B.A., and G. R. BURNELL, C.E. Revised Edition. With a New Appendix on the Qualities of Water. Illustrated. 2s.
35. *THE BLASTING AND QUARRYING OF STONE,* for Building and other Purposes. By Gen. Sir J. BURGOYNE, Bart. 1s. 6d.
43. *TUBULAR, AND OTHER IRON GIRDER BRIDGES,* particularly describing the Britannia and Conway Tubular Bridges. By G. DRYSDALE DEMPSEY, C.E. Fourth Edition. 2s.
44. *FOUNDATIONS AND CONCRETE WORKS,* with Practical Remarks on Footings, Sand, Concrete, Béton, Pile-driving, Caissons, and Cofferdams, &c. By E. DOBSON. Fifth Edition. 1s. 6d.
60. *LAND AND ENGINEERING SURVEYING.* By T. BAKER, C.E. Fifteenth Edition, revised by Professor J. R. YOUNG. 2s.‡
80*. *EMBANKING LANDS FROM THE SEA.* With examples and Particulars of actual Embankments, &c. By J. WIGGINS, F.G.S. 2s.
81. *WATER WORKS,* for the Supply of Cities and Towns. With a Description of the Principal Geological Formations of England as influencing Supplies of Water, &c. By S. HUGHES, C.E. New Edition. 4s.‡
118. *CIVIL ENGINEERING IN NORTH AMERICA,* a Sketch of. By DAVID STEVENSON, F.R.S.E., &c. Plates and Diagrams. 3s.
167. *IRON BRIDGES, GIRDERS, ROOFS, AND OTHER WORKS.* By FRANCIS CAMPIN, C.E. 2s. 6d.‡
197. *ROADS AND STREETS.* By H. LAW, C.E., revised and enlarged by D. K. CLARK, C.E., including pavements of Stone, Wood, Asphalte, &c. 4s. 6d.‡
203. *SANITARY WORK IN THE SMALLER TOWNS AND IN VILLAGES.* By C. SLAGG, A.M.I.C.E. Revised Edition. 3s.‡
212. *GAS-WORKS, THEIR CONSTRUCTION AND ARRANGEMENT;* and the Manufacture and Distribution of Coal Gas. Originally written by SAMUEL HUGHES, C.E. Re-written and enlarged by WILLIAM RICHARDS, C.E. Seventh Edition, with important additions. 5s. 6d.‡
213. *PIONEER ENGINEERING.* A Treatise on the Engineering Operations connected with the Settlement of Waste Lands in New Countries. By EDWARD DOBSON, Assoc. Inst. C.E. 4s. 6d.‡
216. *MATERIALS AND CONSTRUCTION;* A Theoretical and Practical Treatise on the Strains, Designing, and Erection of Works of Construction. By FRANCIS CAMPIN, C.E. Second Edition, revised. 3s.‡
219. *CIVIL ENGINEERING.* By HENRY LAW, M.Inst. C.E. Including HYDRAULIC ENGINEERING by GEO. R. BURNELL, M.Inst. C.E. Seventh Edition, revised, with large additions by D. KINNEAR CLARK, M.Inst. C.E. 6s. 6d., Cloth boards, 7s. 6d.
268. *THE DRAINAGE OF LANDS, TOWNS, & BUILDINGS.* By G. D. DEMPSEY, C.E. Revised, with large Additions on Recent Practice in Drainage Engineering, by D. KINNEAR CLARK, M.I.C.E. Second Edition, Corrected. 4s. 6d.‡ [*Just published.*

☞ *The ‡ indicates that these vols. may be had strongly bound at 6d. extra.*

LONDON : CROSBY LOCKWOOD AND SON,

MECHANICAL ENGINEERING, ETC.

33. *CRANES*, the Construction of, and other Machinery for Raising Heavy Bodies. By JOSEPH GLYNN, F.R.S. Illustrated. 1s. 6d.
34. *THE STEAM ENGINE*. By Dr. LARDNER. Illustrated. 1s. 6d.
59. *STEAM BOILERS*: their Construction and Management. By R. ARMSTRONG, C.E. Illustrated. 1s. 6d.
82. *THE POWER OF WATER*, as applied to drive Flour Mills, and to give motion to Turbines, &c. By JOSEPH GLYNN, F.R.S. 2s.‡
98. *PRACTICAL MECHANISM*, the Elements of; and Machine Tools. By T. BAKER, C.E. With Additions by J. NASMYTH, C.E. 2s. 6d.‡
139. *THE STEAM ENGINE*, a Treatise on the Mathematical Theory of, with Rules and Examples for Practical Men. By T. BAKER, C.E. 1s. 6d.
164. *MODERN WORKSHOP PRACTICE*, as applied to Steam Engines, Bridges, Ship-building, Cranes, &c. By J. G. WINTON. Fourth Edition, much enlarged and carefully revised. 3s. 6d.‡ [*Just published*.
165. *IRON AND HEAT*, exhibiting the Principles concerned in the Construction of Iron Beams, Pillars, and Girders. By J. ARMOUR. 2s. 6d.‡
166. *POWER IN MOTION:* Horse-Power, Toothed-Wheel Gearing, Long and Short Driving Bands, and Angular Forces. By J. ARMOUR. 2s.‡
171. *THE WORKMAN'S MANUAL OF ENGINEERING DRAWING*. By J. MAXTON. 6th Edn. With 7 Plates and 350 Cuts. 3s. 6d.‡
190. *STEAM AND THE STEAM ENGINE*, Stationary and Portable. Being an Extension of the Elementary Treatise on the Steam Engine of MR. JOHN SEWELL. By D. K. CLARK, M.I.C.E. 3s. 6d.‡
200. *FUEL*, its Combustion and Economy. By C. W. WILLIAMS With Recent Practice in the Combustion and Economy of Fuel—Coal, Coke Wood, Peat, Petroleum, &c.—by D. K. CLARK, M.I.C.E. 3s. 6d.‡
202. *LOCOMOTIVE ENGINES*. By G. D. DEMPSEY, C.E.; with large additions by D. KINNEAR CLARK, M.I.C.E. 3s.‡
211. *THE BOILERMAKER'S ASSISTANT* in Drawing, Templating, and Calculating Boiler and Tank Work. By JOHN COURTNEY, Practical Boiler Maker. Edited by D. K. CLARK, C.E. 100 Illustrations. 2s.
217. *SEWING MACHINERY:* Its Construction, History, &c., with full Technical Directions for Adjusting, &c. By J. W. URQUHART, C.E. 2s.‡
223. *MECHANICAL ENGINEERING*. Comprising Metallurgy, Moulding, Casting, Forging, Tools, Workshop Machinery, Manufacture of the Steam Engine, &c. By FRANCIS CAMPIN, C.E. Second Edition. 2s. 6d.‡
236. *DETAILS OF MACHINERY*. Comprising Instructions for the Execution of various Works in Iron. By FRANCIS CAMPIN, C.E. 3s.‡
237. *THE SMITHY AND FORGE;* including the Farrier's Art and Coach Smithing. By W. J. E. CRANE. Illustrated. 2s. 6d.‡
238. *THE SHEET-METAL WORKER'S GUIDE;* a Practical Handbook for Tinsmiths, Coppersmiths, Zincworkers, &c. With 94 Diagrams and Working Patterns. By W. J. E. CRANE. Second Edition, revised. 1s. 6d.
251. *STEAM AND MACHINERY MANAGEMENT:* with Hints on Construction and Selection. By M. POWIS BALE, M.I.M.E. 2s. 6d.‡
254. *THE BOILERMAKER'S READY-RECKONER*. By J. COURTNEY. Edited by D. K. CLARK, C.E. 4s., limp; 5s., half-bound.
255. *LOCOMOTIVE ENGINE-DRIVING*. A Practical Manual for Engineers in charge of Locomotive Engines. By MICHAEL REYNOLDS, M.S.E. Eighth Edition. 3s. 6d., limp; 4s. 6d. cloth boards.
256. *STATIONARY ENGINE-DRIVING*. A Practical Manual for Engineers in charge of Stationary Engines. By MICHAEL REYNOLDS, M.S.E. Third Edition. 3s. 6d. limp; 4s. 6d. cloth boards.
260. *IRON BRIDGES OF MODERATE SPAN:* their Construction and Erection. By HAMILTON W. PENDRED, C.E. 2s.

☞ *The ‡ indicates that these vols. may be had strongly bound at 6d. extra.*

7, STATIONERS' HALL COURT, LUDGATE HILL, E.C.

MINING, METALLURGY, ETC.

4. *MINERALOGY*, Rudiments of; a concise View of the General Properties of Minerals. By A. RAMSAY, F.G.S., F.R.G.S., &c. Third Edition, revised and enlarged. Illustrated. 3s. 6d.‡
117. *SUBTERRANEOUS SURVEYING*, with and without the Magnetic Needle. By T. FENWICK and T. BAKER, C.E. Illustrated. 2s. 6d.‡
135. *ELECTRO-METALLURGY*; Practically Treated. By ALEXANDER WATT. Ninth Edition, enlarged and revised, with additional Illustrations, and including the most recent Processes. 3s. 6d.‡
172. *MINING TOOLS*, Manual of. For the Use of Mine Managers, Agents, Students, &c. By WILLIAM MORGANS. 2s. 6d.
172*. *MINING TOOLS, ATLAS* of Engravings to Illustrate the above, containing 235 Illustrations, drawn to Scale. 4to. 4s. 6d.
176. *METALLURGY OF IRON*. Containing History of Iron Manufacture, Methods of Assay, and Analyses of Iron Ores, Processes of Manufacture of Iron and Steel, &c. By H. BAUERMAN, F.G.S. Sixth Edition, revised and enlarged. 5s.‡ [*Just published*.
180. *COAL AND COAL MINING*. By the late Sir WARINGTON W. SMYTH, M.A., F.R.S. Seventh Edition, revised. 3s. 6d.‡ [*Just published*.
195. *THE MINERAL SURVEYOR AND VALUER'S COMPLETE GUIDE*. By W. LINTERN, M.E. Third Edition, including Magnetic and Angular Surveying. With Four Plates. 3s. 6d.‡
214. *SLATE AND SLATE QUARRYING*, Scientific, Practical, and Commercial. By D. C. DAVIES, F.G.S., Mining Engineer, &c. 3s.‡
264. *A FIRST BOOK OF MINING AND QUARRYING*, with the Sciences connected therewith, for Primary Schools and Self Instruction. By J. H. COLLINS, F.G.S. Second Edition, with additions. 1s. 6d.

ARCHITECTURE, BUILDING, ETC.

16. *ARCHITECTURE—ORDERS*—The Orders and their Æsthetic Principles. By W. H. LEEDS. Illustrated. 1s. 6d.
17. *ARCHITECTURE—STYLES*—The History and Description of the Styles of Architecture of Various Countries, from the Earliest to the Present Period. By T. TALBOT BURY, F.R.I.B.A., &c. Illustrated. 2s.
*⁎*ORDERS AND STYLES OF ARCHITECTURE, *in One Vol*., 3s. 6d.
18. *ARCHITECTURE—DESIGN*—The Principles of Design in Architecture, as deducible from Nature and exemplified in the Works of the Greek and Gothic Architects. By E. L. GARBETT, Architect. Illustrated. 2s. 6d.
⁎ *The three preceding Works, in One handsome Vol., half bound, entitled* "MODERN ARCHITECTURE," *price* 6s.
22. *THE ART OF BUILDING*, Rudiments of. General Principles of Construction, Materials used in Building, Strength and Use of Materials, Working Drawings, Specifications, and Estimates. By E. DOBSON, 2s.‡
25. *MASONRY AND STONECUTTING*: Rudimentary Treatise on the Principles of Masonic Projection and their application to Construction. By EDWARD DOBSON, M.R.I.B.A., &c. 2s. 6d.‡
42. *COTTAGE BUILDING*. By C. BRUCE ALLEN, Architect. Tenth Edition, revised and enlarged. With a Chapter on Economic Cottages for Allotments, by EDWARD E. ALLEN, C.E. 2s.
45. *LIMES, CEMENTS, MORTARS, CONCRETES, MASTICS*, PLASTERING, &c. By G. R. BURNELL, C.E. Thirteenth Edition. 1s. 6d.
57. *WARMING AND VENTILATION*. An Exposition of the General Principles as applied to Domestic and Public Buildings, Mines, Lighthouses, Ships, &c. By C. TOMLINSON, F.R.S., &c. Illustrated. 3s.
111. *ARCHES, PIERS, BUTTRESSES, &c.*: Experimental Essays on the Principles of Construction. By W. BLAND. Illustrated. 1s. 6d.

☞ *The* ‡ *indicates that these vols. may be had strongly bound at* 6d. *extra*.

Architecture, Building, etc., *continued.*

116. *THE ACOUSTICS OF PUBLIC BUILDINGS;* or, The Principles of the Science of Sound applied to the purposes of the Architect and Builder. By T. ROGER SMITH, M.R.I.B.A., Architect. Illustrated. 1s. 6d.
127. *ARCHITECTURAL MODELLING IN PAPER,* the Art of. By T. A. RICHARDSON, Architect. Illustrated. 1s. 6d.
128. *VITRUVIUS—THE ARCHITECTURE OF MARCUS VITRUVIUS POLLO.* In Ten Books. Translated from the Latin by JOSEPH GWILT, F.S.A., F.R.A.S. With 23 Plates. 5s.
130. *GRECIAN ARCHITECTURE,* An Inquiry into the Principles of Beauty in; with an Historical View of the Rise and Progress of the Art in Greece. By the EARL OF ABERDEEN. 1s.

⁎ *The two preceding Works in One handsome Vol., half bound, entitled* "ANCIENT ARCHITECTURE," *price 6s.*

132. *THE ERECTION OF DWELLING-HOUSES.* Illustrated by a Perspective View, Plans, Elevations, and Sections of a pair of Semi-detached Villas, with the Specification, Quantities, and Estimates, &c. By S. H. BROOKS. New Edition, with Plates. 2s. 6d.‡
156. *QUANTITIES & MEASUREMENTS* in Bricklayers', Masons', Plasterers', Plumbers', Painters', Paperhangers', Gilders', Smiths', Carpenters' and Joiners' Work. By A. C. BEATON, Surveyor. New Edition. 1s. 6d.
175. *LOCKWOOD'S BUILDER'S PRICE BOOK FOR* 1891. A Comprehensive Handbook of the Latest Prices and Data for Builders, Architects, Engineers, and Contractors. Re-constructed, Re-written, and greatly Enlarged. By FRANCIS T. W. MILLER, A.R.I.B.A. 650 pages. 3s. 6d.; cloth boards, 4s. [*Just Published.*
182. *CARPENTRY AND JOINERY*—THE ELEMENTARY PRINCIPLES OF CARPENTRY. Chiefly composed from the Standard Work of THOMAS TREDGOLD, C.E. With a TREATISE ON JOINERY by E. WYNDHAM TARN, M.A. Fifth Edition, Revised. 3s. 6d.‡
182*. *CARPENTRY AND JOINERY. ATLAS* of 35 Plates to accompany the above. With Descriptive Letterpress. 4to. 6s.
185. *THE COMPLETE MEASURER;* the Measurement of Boards, Glass, &c.; Unequal-sided, Square-sided, Octagonal-sided, Round Timber and Stone, and Standing Timber, &c. By RICHARD HORTON. Fifth Edition. 4s.; strongly bound in leather, 5s.
187. *HINTS TO YOUNG ARCHITECTS.* By G. WIGHTWICK. New Edition. By G. H. GUILLAUME. Illustrated. 3s. 6d.‡
188. *HOUSE PAINTING, GRAINING, MARBLING, AND SIGN WRITING:* with a Course of Elementary Drawing for House-Painters, Sign-Writers, &c., and a Collection of Useful Receipts. By ELLIS A. DAVIDSON. Sixth Edition. With Coloured Plates. 5s. cloth limp; 6s. cloth boards.
189. *THE RUDIMENTS OF PRACTICAL BRICKLAYING.* In Six Sections: General Principles; Arch Drawing, Cutting, and Setting; Pointing; Paving, Tiling, Materials; Slating and Plastering; Practical Geometry, Mensuration, &c. By ADAM HAMMOND. Seventh Edition. 1s. 6d.
191. *PLUMBING.* A Text-Book to the Practice of the Art or Craft of the Plumber. With Chapters upon House Drainage and Ventilation. Fifth Edition. With 380 Illustrations. By W. P. BUCHAN. 3s. 6d.‡
192. *THE TIMBER IMPORTER'S, TIMBER MERCHANT'S,* and BUILDER'S STANDARD GUIDE. By R. E. GRANDY. 2s.
206. *A BOOK ON BUILDING, Civil and Ecclesiastical,* including CHURCH RESTORATION. With the Theory of Domes and the Great Pyramid, &c. By Sir EDMUND BECKETT, Bart., LL.D., Q.C., F.R.A.S. 4s. 6d.‡
226. *THE JOINTS MADE AND USED BY BUILDERS* in the Construction of various kinds of Engineering and Architectural Works. By WYVILL J. CHRISTY, Architect. With upwards of 160 Engravings on Wood. 3s.‡
228. *THE CONSTRUCTION OF ROOFS OF WOOD AND IRON.* By E. WYNDHAM TARN, M.A., Architect. Second Edition, revised. 1s. 6d.

☞ The ‡ *indicates that these vols. may be had strongly bound at 6d. extra.*

7, STATIONERS' HALL COURT, LUDGATE HILL, E.C.

Architecture, Building, etc., *continued.*

229. *ELEMENTARY DECORATION:* as applied to the Interior and Exterior Decoration of Dwelling-Houses, &c. By J. W. Facey. 2s.

257. *PRACTICAL HOUSE DECORATION.* A Guide to the Art of Ornamental Painting. By James W. Facey. 2s. 6d.

*** *The two preceding Works, in One handsome Vol., half-bound, entitled* "House Decoration, Elementary and Practical," *price* 5s.

230. *HANDRAILING.* Showing New and Simple Methods for finding the Pitch of the Plank, Drawing the Moulds, Bevelling, Jointing-up, and Squaring the Wreath. By George Collings. Second Edition, Revised including A Treatise on Stairbuilding. Plates and Diagrams. 2s. 6d.

247. *BUILDING ESTATES:* a Rudimentary Treatise on the Development, Sale, Purchase, and General Management of Building Land. By Fowler Maitland, Surveyor. Second Edition, revised. 2s.

248. *PORTLAND CEMENT FOR USERS.* By Henry Faija, Assoc. M. Inst. C.E. Third Edition, corrected. Illustrated. 2s.

252. *BRICKWORK:* a Practical Treatise, embodying the General and Higher Principles of Bricklaying, Cutting and Setting, &c. By F. Walker. Second Edition, Revised and Enlarged. 1s. 6d.

23. *THE PRACTICAL BRICK AND TILE BOOK.* Comprising:
189. Brick and Tile Making, by E. Dobson, A.I.C.E.; Practical Bricklaying, by A. Hammond; Brickcutting and Setting, by A. Hammond. 534 pp. with 270 Illustrations. 6s. Strongly half-bound.
265.

253. *THE TIMBER MERCHANT'S, SAW-MILLER'S, AND IMPORTER'S FREIGHT-BOOK AND ASSISTANT.* By Wm. Richardson. With a Chapter on Speeds of Saw-Mill Machinery, &c. By M. Powis Bale, A.M.Inst.C.E. 3s.‡

258. *CIRCULAR WORK IN CARPENTRY AND JOINERY.* A Practical Treatise on Circular Work of Single and Double Curvature. By George Collings. Second Edition, 2s. 6d.

259. *GAS FITTING:* A Practical Handbook treating of every Description of Gas Laying and Fitting. By John Black. With 122 Illustrations. 2s. 6d.‡

261. *SHORING AND ITS APPLICATION:* A Handbook for the Use of Students. By George H. Blagrove. 1s. 6d. [*Just published.*

265. *THE ART OF PRACTICAL BRICK CUTTING & SETTING.* By Adam Hammond. With 90 Engravings. 1s. 6d. [*Just published.*

267. *THE SCIENCE OF BUILDING:* An Elementary Treatise on the Principles of Construction. Adapted to the Requirements of Architectural Students. By E. Wyndham Tarn, M.A. Lond. Third Edition, Revised and Enlarged. With 59 Wood Engravings. 3s. 6d.‡ [*Just published.*

271. *VENTILATION:* a Text-book to the Practice of the Art of Ventilating Buildings, with a Supplementary Chapter upon Air Testing. By William Paton Buchan, R.P., Sanitary and Ventilating Engineer, Author of "Plumbing," &c. 3s. 6d.‡ [*Just published.*

SHIPBUILDING, NAVIGATION, MARINE ENGINEERING, ETC.

51. *NAVAL ARCHITECTURE.* An Exposition of the Elementary Principles of the Science, and their Practical Application to Naval Construction. By J. Peake. Fifth Edition, with Plates and Diagrams. 3s. 6d.‡

53*. *SHIPS FOR OCEAN & RIVER SERVICE,* Elementary and Practical Principles of the Construction of. By H. A. Sommerfeldt. 1s. 6d.

53**. *AN ATLAS OF ENGRAVINGS* to Illustrate the above. Twelve large folding plates. Royal 4to, cloth. 7s. 6d.

54. *MASTING, MAST-MAKING, AND RIGGING OF SHIPS,* Also Tables of Spars, Rigging, Blocks; Chain, Wire, and Hemp Ropes, &c., relative to every class of vessels. By Robert Kipping, N.A. 2s.

☞ *The* ‡ *indicates that these vols. may be had strongly bound at 6d. extra.*

LONDON: CROSBY LOCKWOOD AND SON,

WEALE'S RUDIMENTARY SERIES.

Shipbuilding, Navigation, Marine Engineering, etc., *cont.*

54*. *IRON SHIP-BUILDING*. With Practical Examples and Details. By John Grantham, C.E. Fifth Edition. 4s.

55. *THE SAILOR'S SEA BOOK*: a Rudimentary Treatise on Navigation. By James Greenwood, B.A. With numerous Woodcuts and Coloured Plates. New and enlarged edition. By W. H. Rosser. 2s. 6d.‡

80. *MARINE ENGINES AND STEAM VESSELS*. By Robert Murray, C.E. Eighth Edition, thoroughly Revised, with Additions by the Author and by George Carlisle, C.E. 4s. 6d. limp; 5s. cloth boards.

83*bis*. *THE FORMS OF SHIPS AND BOATS*. By W. Bland. Seventh Edition, Revised, with numerous Illustrations and Models. 1s. 6d.

99. *NAVIGATION AND NAUTICAL ASTRONOMY*, in Theory and Practice. By Prof. J. R. Young. New Edition. 2s. 6d.

106. *SHIPS' ANCHORS*, a Treatise on. By G. Cotsell, N.A. 1s. 6d.

149. *SAILS AND SAIL-MAKING*. With Draughting, and the Centre of Effort of the Sails; Weights and Sizes of Ropes; Masting, Rigging, and Sails of Steam Vessels, &c. 12th Edition. By R. Kipping, N.A. 2s. 6d.‡

155. *ENGINEER'S GUIDE TO THE ROYAL & MERCANTILE NAVIES*. By a Practical Engineer. Revised by D. F. M'Carthy. 3s.

55 & 204. *PRACTICAL NAVIGATION*. Consisting of The Sailor's Sea-Book. By James Greenwood and W. H. Rosser. Together with the requisite Mathematical and Nautical Tables for the Working of the Problems. By H. Law, C.E., and Prof. J. R. Young. 7s. Half-bound.

AGRICULTURE, GARDENING, ETC.

61*. *A COMPLETE READY RECKONER FOR THE ADMEASUREMENT OF LAND*, &c. By A. Arman. Third Edition, revised and extended by C. Norris, Surveyor, Valuer, &c. 2s.

131. *MILLER'S, CORN MERCHANT'S, AND FARMER'S READY RECKONER*. Second Edition, with a Price List of Modern Flour-Mill Machinery, by W. S. Hutton, C.E. 2s.

140. *SOILS, MANURES, AND CROPS*. (Vol. 1. Outlines of Modern Farming.) By R. Scott Burn. Woodcuts. 2s.

141. *FARMING & FARMING ECONOMY*, Notes, Historical and Practical, on. (Vol. 2. Outlines of Modern Farming.) By R. Scott Burn. 3s.

142. *STOCK; CATTLE, SHEEP, AND HORSES*. (Vol. 3. Outlines of Modern Farming.) By R. Scott Burn. Woodcuts. 2s. 6d.

145. *DAIRY, PIGS, AND POULTRY*, Management of the. By R. Scott Burn. (Vol. 4. Outlines of Modern Farming.) 2s.

146. *UTILIZATION OF SEWAGE, IRRIGATION, AND RECLAMATION OF WASTE LAND*. (Vol. 5. Outlines of Modern Farming.) By R. Scott Burn. Woodcuts. 2s. 6d.

⁎ Nos. 140-1-2-5-6, *in One Vol., handsomely half-bound, entitled* "Outlines of Modern Farming." By Robert Scott Burn. *Price* 12s.

177. *FRUIT TREES*, The Scientific and Profitable Culture of. From the French of Du Breuil. Revised by Geo. Glenny. 187 Woodcuts. 3s. 6d.‡

198. *SHEEP; THE HISTORY, STRUCTURE, ECONOMY, AND DISEASES OF*. By W. C. Spooner, M.R.V.C., &c. Fifth Edition, enlarged, including Specimens of New and Improved Breeds. 3s. 6d.‡

201. *KITCHEN GARDENING MADE EASY*. By George M. F. Glenny. Illustrated. 1s. 6d.‡

207. *OUTLINES OF FARM MANAGEMENT, and the Organization of Farm Labour*. By R. Scott Burn. 2s. 6d.‡

208. *OUTLINES OF LANDED ESTATES MANAGEMENT*. By R. Scott Burn. 2s. 6d.

⁎ Nos. 207 & 208 *in One Vol., handsomely half-bound, entitled* "Outlines of Landed Estates and Farm Management." By R. Scott Burn. *Price* 6s.

☞ *The ‡ indicates that these vols. may be had strongly bound at 6d. extra.*

7, STATIONERS' HALL COURT, LUDGATE HILL, E.C.

Agriculture, Gardening, etc., *continued.*

209. *THE TREE PLANTER AND PLANT PROPAGATOR.* A Practical Manual on the Propagation of Forest Trees, Fruit Trees, Flowering Shrubs, Flowering Plants, &c. By SAMUEL WOOD. 2s.

210. *THE TREE PRUNER.* A Practical Manual on the Pruning of Fruit Trees, including also their Training and Renovation; also the Pruning of Shrubs, Climbers, and Flowering Plants. By SAMUEL WOOD. 1s. 6d.

*** *Nos.* 209 & 210 *in One Vol., handsomely half-bound, entitled* "THE TREE PLANTER, PROPAGATOR, AND PRUNER." By SAMUEL WOOD. *Price* 3s. 6d.

218. *THE HAY AND STRAW MEASURER:* Being New Tables for the Use of Auctioneers, Valuers, Farmers, Hay and Straw Dealers, &c. By JOHN STEELE. Fourth Edition. 2s.

222. *SUBURBAN FARMING.* The Laying-out and Cultivation of Farms, adapted to the Produce of Milk, Butter, and Cheese, Eggs, Poultry, and Pigs. By Prof. JOHN DONALDSON and R. SCOTT BURN. 3s. 6d.‡

231. *THE ART OF GRAFTING AND BUDDING.* By CHARLES BALTET. With Illustrations. 2s. 6d.‡

232. *COTTAGE GARDENING;* or, Flowers, Fruits, and Vegetables for Small Gardens. By E. HOBDAY. 1s. 6d.

233. *GARDEN RECEIPTS.* Edited by CHARLES W. QUIN. 1s. 6d.

234. *MARKET AND KITCHEN GARDENING.* By C. W. SHAW, late Editor of "Gardening Illustrated." 3s.‡ [*Just published.*

239. *DRAINING AND EMBANKING.* A Practical Treatise, embodying the most recent experience in the Application of Improved Methods. By JOHN SCOTT, late Professor of Agriculture and Rural Economy at the Royal Agricultural College, Cirencester. With 68 Illustrations. 1s. 6d.

240. *IRRIGATION AND WATER SUPPLY.* A Treatise on Water Meadows, Sewage Irrigation, and Warping; the Construction of Wells, Ponds, and Reservoirs, &c. By Prof. JOHN SCOTT. With 34 Illus. 1s. 6d.

241. *FARM ROADS, FENCES, AND GATES.* A Practical Treatise on the Roads, Tramways, and Waterways of the Farm; the Principles of Enclosures; and the different kinds of Fences, Gates, and Stiles. By Professor JOHN SCOTT. With 75 Illustrations. 1s. 6d.

242. *FARM BUILDINGS.* A Practical Treatise on the Buildings necessary for various kinds of Farms, their Arrangement and Construction, with Plans and Estimates. By Prof. JOHN SCOTT. With 105 Illus. 2s.

243. *BARN IMPLEMENTS AND MACHINES.* A Practical Treatise on the Application of Power to the Operations of Agriculture; and on various Machines used in the Threshing-barn, in the Stock-yard, and in the Dairy, &c. By Prof. J. SCOTT. With 123 Illustrations. 2s.

244. *FIELD IMPLEMENTS AND MACHINES.* A Practical Treatise on the Varieties now in use, with Principles and Details of Construction, their Points of Excellence, and Management. By Professor JOHN SCOTT. With 138 Illustrations. 2s.

245. *AGRICULTURAL SURVEYING.* A Practical Treatise on Land Surveying, Levelling, and Setting-out; and on Measuring and Estimating Quantities, Weights, and Values of Materials, Produce, Stock, &c. By Prof. JOHN SCOTT. With 62 Illustrations. 1s. 6d.

*** *Nos.* 239 *to* 245 *in One Vol., handsomely half-bound, entitled* "THE COMPLETE TEXT-BOOK OF FARM ENGINEERING." By Professor JOHN SCOTT. *Price* 12s.

250. *MEAT PRODUCTION.* A Manual for Producers, Distributors, &c. By JOHN EWART. 2s. 6d.‡

266. *BOOK-KEEPING FOR FARMERS & ESTATE OWNERS.* By J. M. WOODMAN, Chartered Accountant. 2s. 6d. cloth limp; 3s. 6d. cloth boards. [*Just published.*

☞ *The* ‡ *indicates that these vols. may be had strongly bound at* 6d. *extra.*

LONDON: CROSBY LOCKWOOD AND SON,

MATHEMATICS, ARITHMETIC, ETC.

32. *MATHEMATICAL INSTRUMENTS*, a Treatise on; Their Construction, Adjustment, Testing, and Use concisely Explained. By J. F. HEATHER, M.A. Fourteenth Edition, revised, with additions, by A. T. WALMISLEY, M.I.C.E., Fellow of the Surveyors' Institution. Original Edition, in 1 vol., Illustrated. 2s.‡ [*Just published.*

⁎⁎ In ordering the above, be careful to say, "Original Edition" (*No.* 32), to distinguish it from the Enlarged Edition in 3 vols. (*Nos.* 168-9-70.)

76. *DESCRIPTIVE GEOMETRY*, an Elementary Treatise on; with a Theory of Shadows and of Perspective, extracted from the French of G. MONGE. To which is added, a description of the Principles and Practice of Isometrical Projection. By J. F. HEATHER, M.A. With 14 Plates. 2s.

178. *PRACTICAL PLANE GEOMETRY:* giving the Simplest Modes of Constructing Figures contained in one Plane and Geometrical Construction of the Ground. By J. F. HEATHER, M.A. With 215 Woodcuts. 2s.

83. *COMMERCIAL BOOK-KEEPING*. With Commercial Phrases and Forms in English, French, Italian, and German. By JAMES HADDON, M.A., Arithmetical Master of King's College School, London. 1s. 6d.

84. *ARITHMETIC*, a Rudimentary Treatise on: with full Explanations of its Theoretical Principles, and numerous Examples for Practice. By Professor J. R. YOUNG. Eleventh Edition. 1s. 6d.

84*. A KEY to the above, containing Solutions in full to the Exercises, together with Comments, Explanations, and Improved Processes, for the Use of Teachers and Unassisted Learners. By J. R. YOUNG. 1s. 6d.

85. *EQUATIONAL ARITHMETIC*, applied to Questions of Interest, Annuities, Life Assurance, and General Commerce; with various Tables by which all Calculations may be greatly facilitated. By W. HIPSLEY. 2s.

86. *ALGEBRA*, the Elements of. By JAMES HADDON, M.A. With Appendix, containing miscellaneous Investigations, and a Collection of Problems in various parts of Algebra. 2s.

86*. A KEY AND COMPANION to the above Book, forming an extensive repository of Solved Examples and Problems in Illustration of the various Expedients necessary in Algebraical Operations. By J. R. YOUNG. 1s. 6d.

88. *EUCLID*, THE ELEMENTS OF: with many additional Propositions
89. and Explanatory Notes: to which is prefixed, an Introductory Essay on Logic. By HENRY LAW, C.E. 2s. 6d.‡

⁎⁎ *Sold also separately, viz.:—*

88. EUCLID, The First Three Books. By HENRY LAW, C.E. 1s. 6d.
89. EUCLID, Books 4, 5, 6, 11, 12. By HENRY LAW, C.E. 1s. 6d.

90. *ANALYTICAL GEOMETRY AND CONIC SECTIONS*, By JAMES HANN. A New Edition, by Professor J. R. YOUNG. 2s.‡

91. *PLANE TRIGONOMETRY*, the Elements of. By JAMES HANN, formerly Mathematical Master of King's College, London. 1s. 6d.

92. *SPHERICAL TRIGONOMETRY*, the Elements of. By JAMES HANN. Revised by CHARLES H. DOWLING, C.E. 1s.
⁎⁎ *Or with "The Elements of Plane Trigonometry," in One Volume,* 2s. 6d.

93. *MENSURATION AND MEASURING*. With the Mensuration and Levelling of Land for the Purposes of Modern Engineering. By T. BAKER, C.E. New Edition by E. NUGENT, C.E. Illustrated. 1s. 6d.

101. *DIFFERENTIAL CALCULUS*, Elements of the. By W. S. B. WOOLHOUSE, F.R.A.S., &c. 1s. 6d.

102. *INTEGRAL CALCULUS*, Rudimentary Treatise on the. By HOMERSHAM COX, B.A. Illustrated. 1s.

136. *ARITHMETIC*, Rudimentary, for the Use of Schools and Self-Instruction. By JAMES HADDON, M.A. Revised by A. ARMAN. 1s. 6d.
137. A KEY TO HADDON'S RUDIMENTARY ARITHMETIC. By A. ARMAN. 1s. 6d.

☞ *The ‡ indicates that these vols. may be had strongly bound at* 6d. *extra.*

7, STATIONERS' HALL COURT, LUDGATE HILL, E.C.

Mathematics, Arithmetic, etc., *continued.*

168. *DRAWING AND MEASURING INSTRUMENTS.* Including—I. Instruments employed in Geometrical and Mechanical Drawing, and in the Construction, Copying, and Measurement of Maps and Plans. II. Instruments used for the purposes of Accurate Measurement, and for Arithmetical Computations. By J. F. HEATHER, M.A. Illustrated. 1s. 6d.
169. *OPTICAL INSTRUMENTS.* Including (more especially) Telescopes, Microscopes, and Apparatus for producing copies of Maps and Plans by Photography. By J. F. HEATHER, M.A. Illustrated. 1s. 6d.
170. *SURVEYING AND ASTRONOMICAL INSTRUMENTS.* Including—I. Instruments Used for Determining the Geometrical Features of a portion of Ground. II. Instruments Employed in Astronomical Observations. By J. F. HEATHER, M.A. Illustrated. 1s. 6d.

* *The above three volumes form an enlargement of the Author's original work "Mathematical Instruments." (See No. 32 in the Series.)*

168.⎫
169.⎬ *MATHEMATICAL INSTRUMENTS.* By J. F. HEATHER, M.A. Enlarged Edition, for the most part entirely re-written. The 3 Parts as
170.⎭ above, in One thick Volume. With numerous Illustrations. 4s. 6d.‡

158. *THE SLIDE RULE, AND HOW TO USE IT;* containing full, easy, and simple Instructions to perform all Business Calculations with unexampled rapidity and accuracy. By CHARLES HOARE, C.E. Fifth Edition. With a Slide Rule in tuck of cover. 2s. 6d.‡
196. *THEORY OF COMPOUND INTEREST AND ANNUITIES;* with Tables of Logarithms for the more Difficult Computations of Interest, Discount, Annuities, &c. By FÉDOR THOMAN. 4s.‡
199. *THE COMPENDIOUS CALCULATOR;* or, Easy and Concise Methods of Performing the various Arithmetical Operations required in Commercial and Business Transactions; together with Useful Tables. By D. O'GORMAN. Twenty-seventh Edition, carefully revised by C. NORRIS. 2s. 6d., cloth limp; 3s. 6d., strongly half-bound in leather.
204. *MATHEMATICAL TABLES,* for Trigonometrical, Astronomical, and Nautical Calculations; to which is prefixed a Treatise on Logarithms. By HENRY LAW, C.E. Together with a Series of Tables for Navigation and Nautical Astronomy. By Prof. J. R. YOUNG. New Edition. 4s.
204*. *LOGARITHMS.* With Mathematical Tables for Trigonometrical, Astronomical, and Nautical Calculations. By HENRY LAW, M.Inst.C.E. New and Revised Edition. (Forming part of the above Work). 3s.
221. *MEASURES, WEIGHTS, AND MONEYS OF ALL NATIONS,* and an Analysis of the Christian, Hebrew, and Mahometan Calendars. By W. S. B. WOOLHOUSE, F.R.A.S., F.S.S. Seventh Edition. 2s. 6d.‡
227. *MATHEMATICS AS APPLIED TO THE CONSTRUCTIVE ARTS.* Illustrating the various processes of Mathematical Investigation, by means of Arithmetical and Simple Algebraical Equations and Practical Examples. By FRANCIS CAMPIN, C.E. Second Edition. 3s.‡

PHYSICAL SCIENCE, NATURAL PHILOSOPHY, ETC.

1. *CHEMISTRY.* By Professor GEORGE FOWNES, F.R.S. With an Appendix on the Application of Chemistry to Agriculture. 1s.
2. *NATURAL PHILOSOPHY,* Introduction to the Study of. By C. TOMLINSON. Woodcuts. 1s. 6d.
6. *MECHANICS,* Rudimentary Treatise on. By CHARLES TOMLINSON. Illustrated. 1s. 6d.
7. *ELECTRICITY;* showing the General Principles of Electrical Science, and the purposes to which it has been applied. By Sir W. SNOW HARRIS, F.R.S., &c. With Additions by R. SABINE, C.E., F.S.A. 1s. 6d.
7*. *GALVANISM.* By Sir W. SNOW HARRIS. New Edition by ROBERT SABINE, C.E., F.S.A. 1s. 6d.
8. *MAGNETISM;* being a concise Exposition of the General Principles of Magnetical Science. By Sir W. SNOW HARRIS. New Edition, revised by H. M. NOAD, Ph.D. With 165 Woodcuts. 3s. 6d.‡

☞ *The ‡ indicates that these vols. may be had strongly bound at 6d. extra.*

LONDON: CROSBY LOCKWOOD AND SON,

Physical Science, Natural Philosophy, etc., *continued*.

11. *THE ELECTRIC TELEGRAPH;* its History and Progress; with Descriptions of some of the Apparatus. By R. SABINE, C.E., F.S.A. 3s.
12. *PNEUMATICS*, including Acoustics and the Phenomena of Wind Currents, for the Use of Beginners. By CHARLES TOMLINSON, F.R.S. Fourth Edition, enlarged. Illustrated. 1s. 6d. [*Just published.*
72. *MANUAL OF THE MOLLUSCA;* a Treatise on Recent and Fossil Shells. By Dr. S. P. WOODWARD, A.L.S. Fourth Edition. With Plates and 300 Woodcuts. 7s. 6d., cloth.
96. *ASTRONOMY.* By the late Rev. ROBERT MAIN, M.A. Third Edition, by WILLIAM THYNNE LYNN, B.A., F.R.A.S. 2s.
97. *STATICS AND DYNAMICS*, the Principles and Practice of; embracing also a clear development of Hydrostatics, Hydrodynamics, and Central Forces. By T. BAKER, C.E. Fourth Edition. 1s. 6d.
173. *PHYSICAL GEOLOGY*, partly based on Major-General PORTLOCK's "Rudiments of Geology." By RALPH TATE, A.L.S., &c. Woodcuts. 2s.
174. *HISTORICAL GEOLOGY*, partly based on Major-General PORTLOCK's "Rudiments." By RALPH TATE, A.L.S., &c. Woodcuts. 2s. 6d.
173 & 174. *RUDIMENTARY TREATISE ON GEOLOGY*, Physical and Historical. Partly based on Major-General PORTLOCK's " Rudiments of Geology." By RALPH TATE, A.L.S., F.G.S., &c. In One Volume. 4s. 6d.‡
183 & 184. *ANIMAL PHYSICS*, Handbook of. By Dr. LARDNER, D.C.L., formerly Professor of Natural Philosophy and Astronomy in University College, Lond. With 520 Illustrations. In One Vol. 7s. 6d., cloth boards.
*** *Sold also in Two Parts, as follows :—*
183. ANIMAL PHYSICS. By Dr. LARDNER. Part I., Chapters I.—VII. 4s.
184. ANIMAL PHYSICS. By Dr. LARDNER. Part II., Chapters VIII.—XVIII. 3s.
269. *LIGHT:* an Introduction to the Science of Optics, for the Use of Students of Architecture, Engineering, and other Applied Sciences. By E. WYNDHAM TARN, M.A. 1s. 6d. [*Just published.*

FINE ARTS.

20. *PERSPECTIVE FOR BEGINNERS.* Adapted to Young Students and Amateurs in Architecture, Painting, &c. By GEORGE PYNE. 2s.
40 *GLASS STAINING, AND THE ART OF PAINTING ON GLASS.* From the German of Dr. GESSERT and EMANUEL OTTO FROMBERG. With an Appendix on THE ART OF ENAMELLING. 2s. 6d.
69. *MUSIC*, A Rudimentary and Practical Treatise on. With numerous Examples. By CHARLES CHILD SPENCER. 2s. 6d.
71. *PIANOFORTE*, The Art of Playing the. With numerous Exercises & Lessons from the Best Masters. By CHARLES CHILD SPENCER. 1s.6d.
69-71. *MUSIC & THE PIANOFORTE.* In one vol. Half bound, 5s.
181. *PAINTING POPULARLY EXPLAINED*, including Fresco, Oil, Mosaic, Water Colour, Water-Glass, Tempera, Encaustic, Miniature, Painting on Ivory, Vellum, Pottery, Enamel, Glass, &c. With Historical Sketches of the Progress of the Art by THOMAS JOHN GULLICK, assisted by JOHN TIMBS, F.S.A. Fifth Edition, revised and enlarged. 5s.‡
186. *A GRAMMAR OF COLOURING*, applied to Decorative Painting and the Arts. By GEORGE FIELD. New Edition, enlarged and adapted to the Use of the Ornamental Painter and Designer. By ELLIS A. DAVIDSON. With two new Coloured Diagrams, &c. 3s.‡
246. *A DICTIONARY OF PAINTERS, AND HANDBOOK FOR PICTURE AMATEURS*; including Methods of Painting, Cleaning, Relining and Restoring, Schools of Painting, &c. With Notes on the Copyists and Imitators of each Master. By PHILIPPE DARYL. 2s. 6d.‡

☞ *The ‡ indicates that these vols. may be had strongly bound at 6d. extra.*

INDUSTRIAL AND USEFUL ARTS.

23. *BRICKS AND TILES*, Rudimentary Treatise on the Manufacture of. By E. DOBSON, M.R.I.B.A. Illustrated, 3s.‡
67. *CLOCKS, WATCHES, AND BELLS*, a Rudimentary Treatise on. By Sir EDMUND BECKETT, LL.D., Q.C. Seventh Edition, revised and enlarged. 4s. 6d. limp; 5s. 6d. cloth boards.
83**. *CONSTRUCTION OF DOOR LOCKS.* Compiled from the Papers of A. C. HOBBS, and Edited by CHARLES TOMLINSON, F.R.S. 2s. 6d.
162. *THE BRASS FOUNDER'S MANUAL;* Instructions for Modelling, Pattern-Making, Moulding, Turning, Filing, Burnishing, Bronzing, &c. With copious Receipts, &c. By WALTER GRAHAM. 2s.‡
205. *THE ART OF LETTER PAINTING MADE EASY.* By J. G. BADENOCH. Illustrated with 12 full-page Engravings of Examples. 1s. 6d.
215. *THE GOLDSMITH'S HANDBOOK*, containing full Instructions for the Alloying and Working of Gold. By GEORGE E. GEE, 3s.‡
225. *THE SILVERSMITH'S HANDBOOK*, containing full Instructions for the Alloying and Working of Silver. By GEORGE E. GEE. 3s.‡

. The two preceding Works, in One handsome Vol., half-bound, entitled "THE GOLDSMITH'S & SILVERSMITH'S COMPLETE HANDBOOK," 7s.

249. *THE HALL-MARKING OF JEWELLERY PRACTICALLY CONSIDERED.* By GEORGE E. GEE. 3s.‡
224. *COACH BUILDING*, A Practical Treatise, Historical and Descriptive. By J. W. BURGESS. 2s. 6d.‡
235. *PRACTICAL ORGAN BUILDING.* By W. E. DICKSON, M.A., Precentor of Ely Cathedral. Illustrated. 2s. 6d.‡
262. *THE ART OF BOOT AND SHOEMAKING.* By JOHN BEDFORD LENO. Numerous Illustrations. Third Edition. 2s.
263. *MECHANICAL DENTISTRY:* A Practical Treatise on the Construction of the Various Kinds of Artificial Dentures, with Formulæ, Tables, Receipts, &c. By CHARLES HUNTER. Third Edition. 3s.‡
270. *WOOD ENGRAVING:* A Practical and Easy Introduction to the Study of the Art. By W. N. BROWN. 1s. 6d.

MISCELLANEOUS VOLUMES.

36. *A DICTIONARY OF TERMS used in ARCHITECTURE, BUILDING, ENGINEERING, MINING, METALLURGY, ARCHÆOLOGY, the FINE ARTS, &c.* By JOHN WEALE. Fifth Edition. Revised by ROBERT HUNT, F.R.S. Illustrated. 5s. limp; 6s. cloth boards.
50. *THE LAW OF CONTRACTS FOR WORKS AND SERVICES.* By DAVID GIBBONS. Third Edition, enlarged. 3s.‡
112. *MANUAL OF DOMESTIC MEDICINE.* By R. GOODING, B.A., M.D. A Family Guide in all Cases of Accident and Emergency. 2s.
112*. *MANAGEMENT OF HEALTH.* A Manual of Home and Personal Hygiene. By the Rev. JAMES BAIRD, B.A. 1s.
150. *LOGIC*, Pure and Applied. By S. H. EMMENS. 1s. 6d.
153. *SELECTIONS FROM LOCKE'S ESSAYS ON THE HUMAN UNDERSTANDING.* With Notes by S. H. EMMENS. 2s.
154. *GENERAL HINTS TO EMIGRANTS.* 2s.
157. *THE EMIGRANT'S GUIDE TO NATAL.* By ROBERT JAMES MANN, F.R.A.S., F.M.S. Second Edition. Map. 2s.
193. *HANDBOOK OF FIELD FORTIFICATION.* By Major W. W. KNOLLYS, F.R.G.S. With 163 Woodcuts. 3s.‡
194. *THE HOUSE MANAGER:* Being a Guide to Housekeeping. Practical Cookery, Pickling and Preserving, Household Work, Dairy Management, &c. By AN OLD HOUSEKEEPER. 3s. 6d.‡
194, *HOUSE BOOK (The).* Comprising :—I. THE HOUSE MANAGER.
112 & By an OLD HOUSEKEEPER. II. DOMESTIC MEDICINE. By R. GOODING, M.D.
112* III. MANAGEMENT OF HEALTH. By J. BAIRD. In One Vol., half-bound, 6s.

☞ *The ‡ indicates that these vols may be had strongly bound at 6d. extra.*

LONDON: CROSBY LOCKWOOD AND SON,

EDUCATIONAL AND CLASSICAL SERIES.

HISTORY.

1. **England, Outlines of the History of;** more especially with reference to the Origin and Progress of the English Constitution. By WILLIAM DOUGLAS HAMILTON, F.S.A., of Her Majesty's Public Record Office. 4th Edition, revised. 5s.; cloth boards, 6s.
5. **Greece, Outlines of the History of;** in connection with the Rise of the Arts and Civilization in Europe. By W. DOUGLAS HAMILTON, of University College, London, and EDWARD LEVIEN, M.A., of Balliol College, Oxford. 2s. 6d.; cloth boards, 3s. 6d.
7. **Rome, Outlines of the History of:** from the Earliest Period to the Christian Era and the Commencement of the Decline of the Empire. By EDWARD LEVIEN, of Balliol College, Oxford. Map, 2s. 6d.; cl. bds. 3s. 6d.
9. **Chronology of History, Art, Literature, and Progress,** from the Creation of the World to the Present Time. The Continuation by W. D. HAMILTON, F.S.A. 3s.; cloth boards, 3s. 6d.
50. **Dates and Events in English History,** for the use of Candidates in Public and Private Examinations. By the Rev. E. RAND. 1s.

ENGLISH LANGUAGE AND MISCELLANEOUS.

11. **Grammar of the English Tongue,** Spoken and Written. With an Introduction to the Study of Comparative Philology. By HYDE CLARKE, D.C.L. Fourth Edition. 1s. 6d.
12. **Dictionary of the English Language,** as Spoken and Written. Containing above 100,000 Words. By HYDE CLARKE, D.C.I. 3s. 6d.; cloth boards, 4s. 6d.; complete with the GRAMMAR, cloth bds., 5s. 6d.
48. **Composition and Punctuation,** familiarly Explained for those who have neglected the Study of Grammar. By JUSTIN BRENAN. 18th Edition. 1s. 6d.
49. **Derivative Spelling-Book:** Giving the Origin of Every Word from the Greek, Latin, Saxon, German, Teutonic, Dutch, French, Spanish, and other Languages; with their present Acceptation and Pronunciation. By J. ROWBOTHAM, F.R.A.S. Improved Edition. 1s. 6d.
51. **The Art of Extempore Speaking:** Hints for the Pulpit, the Senate, and the Bar. By M. BAUTAIN, Vicar-General and Professor at the Sorbonne. Translated from the French. 8th Edition, carefully corrected. 2s. 6d.
54. **Analytical Chemistry,** Qualitative and Quantitative, a Course of. To which is prefixed, a Brief Treatise upon Modern Chemical Nomenclature and Notation. By WM. W. PINK and GEORGE E. WEBSTER. 2s.

THE SCHOOL MANAGERS' SERIES OF READING BOOKS,

Edited by the Rev. A. R. GRANT, Rector of Hitcham, and Honorary Canon of Ely; formerly H.M. Inspector of Schools.

INTRODUCTORY PRIMER, 3d.

	s. d.		s. d.
FIRST STANDARD	0 6	FOURTH STANDARD	1 2
SECOND ,,	0 10	FIFTH ,,	1 6
THIRD ,,	1 0	SIXTH ,,	1 6

LESSONS FROM THE BIBLE. Part I. Old Testament. 1s.
LESSONS FROM THE BIBLE. Part II. New Testament, to which is added THE GEOGRAPHY OF THE BIBLE, for very young Children. By Rev. C. THORNTON FORSTER. 1s. 2d. *** Or the Two Parts in One Volume. 2s.

7, STATIONERS' HALL COURT, LUDGATE HILL, E.C.

FRENCH.

24. **French Grammar.** With Complete and Concise Rules on the Genders of French Nouns. By G. L. STRAUSS, Ph.D. 1s. 6d.
25. **French-English Dictionary.** Comprising a large number of New Terms used in Engineering, Mining, &c. By ALFRED ELWES. 1s. 6d
26. **English-French Dictionary.** By ALFRED ELWES. 2s.
25,26. **French Dictionary** (as above). Complete, in One Vol., 3s.; cloth boards, 3s. 6d. *⁎* Or with the GRAMMAR, cloth boards, 4s. 6d.
47. **French and English Phrase Book:** containing Introductory Lessons, with Translations, several Vocabularies of Words, a Collection of suitable Phrases, and Easy Familiar Dialogues. 1s. 6d.

GERMAN.

39. **German Grammar.** Adapted for English Students, from Heyse's Theoretical and Practical Grammar, by Dr. G. L. STRAUSS. 1s. 6d.
40. **German Reader:** A Series of Extracts, carefully culled from the most approved Authors of Germany; with Notes, Philological and Explanatory. By G. L. STRAUSS, Ph.D. 1s.
41-43. **German Triglot Dictionary.** By N. E. S. A. HAMILTON. In Three Parts. Part I. German-French-English. Part II. English-German-French. Part III. French-German-English. 3s., or cloth boards, 4s.
41-43 & 39. **German Triglot Dictionary** (as above), together with German Grammar (No. 39), in One Volume, cloth boards, 5s.

ITALIAN.

27. **Italian Grammar,** arranged in Twenty Lessons, with a Course of Exercises. By ALFRED ELWES. 1s. 6d.
28. **Italian Triglot Dictionary,** wherein the Genders of all the Italian and French Nouns are carefully noted down. By ALFRED ELWES. Vol. 1. Italian-English-French. 2s. 6d.
30. **Italian Triglot Dictionary.** By A. ELWES. Vol. 2. English-French-Italian. 2s. 6d.
32. **Italian Triglot Dictionary.** By ALFRED ELWES. Vol. 3. French-Italian-English. 2s. 6d.
28,30, **Italian Triglot Dictionary** (as above). In One Vol., 7s. 6d. 32. Cloth boards.

SPANISH AND PORTUGUESE.

34. **Spanish Grammar,** in a Simple and Practical Form. With a Course of Exercises. By ALFRED ELWES. 1s. 6d.
35. **Spanish-English and English-Spanish Dictionary.** Including a large number of Technical Terms used in Mining, Engineering, &c. with the proper Accents and the Gender of every Noun. By ALFRED ELWES 4s.; cloth boards, 5s. *⁎* Or with the GRAMMAR, cloth boards, 6s.
55. **Portuguese Grammar,** in a Simple and Practical Form. With a Course of Exercises. By ALFRED ELWES. 1s. 6d.
56. **Portuguese-English and English-Portuguese Dictionary.** Including a large number of Technical Terms used in Mining, Engineering, &c., with the proper Accents and the Gender of every Noun. By ALFRED ELWES. Second Edition, Revised, 5s.; cloth boards, 6s. *⁎* Or with the GRAMMAR, cloth boards, 7s.

HEBREW.

46*. **Hebrew Grammar.** By Dr. BRESSLAU. 1s. 6d.
44. **Hebrew and English Dictionary,** Biblical and Rabbinical; containing the Hebrew and Chaldee Roots of the Old Testament Post-Rabbinical Writings. By Dr. BRESSLAU. 6s.
46. **English and Hebrew Dictionary.** By Dr. BRESSLAU. 3s.
44,46. **Hebrew Dictionary** (as above), in Two Vols., complete, with 46*. the GRAMMAR, cloth boards, 12s.

LONDON: CROSBY LOCKWOOD AND SON,

LATIN.

19. **Latin Grammar.** Containing the Inflections and Elementary Principles of Translation and Construction. By the Rev. THOMAS GOODWIN, M.A., Head Master of the Greenwich Proprietary School. 1s. 6d.
20. **Latin-English Dictionary.** By the Rev. THOMAS GOODWIN, M.A. 2s.
22. **English-Latin Dictionary;** together with an Appendix of French and Italian Words which have their origin from the Latin. By the Rev. THOMAS GOODWIN, M.A. 1s. 6d.
20,22. **Latin Dictionary** (as above). Complete in One Vol., 3s. 6d. cloth boards, 4s. 6d. *** Or with the GRAMMAR, cloth boards, 5s. 6d.

LATIN CLASSICS. With Explanatory Notes in English.
1. **Latin Delectus.** Containing Extracts from Classical Authors, with Genealogical Vocabularies and Explanatory Notes, by H. YOUNG. 1s. 6d.
2. **Cæsaris Commentarii de Bello Gallico.** Notes, and a Geographical Register for the Use of Schools, by H. YOUNG. 2s.
3. **Cornelius Nepos.** With Notes. By H. YOUNG. 1s.
4. **Virgilii Maronis Bucolica et Georgica.** With Notes on the Bucolics by W. RUSHTON, M.A., and on the Georgics by H. YOUNG. 1s. 6d.
5. **Virgilii Maronis Æneis.** With Notes, Critical and Explanatory, by H. YOUNG. New Edition, revised and improved With copious Additional Notes by Rev. T. H. L. LEARY, D.C.L., formerly Scholar of Brasenose College, Oxford. 3s.
5*. ——— Part 1. Books i.—vi., 1s. 6d.
5**. ——— Part 2. Books vii.—xii., 2s.
6. **Horace;** Odes, Epode, and Carmen Sæculare. Notes by H. YOUNG. 1s. 6d.
7. **Horace;** Satires, Epistles, and Ars Poetica. Notes by W. BROWNRIGG SMITH, M.A., F.R.G.S. 1s. 6d.
8. **Sallustii Crispi Catalina et Bellum Jugurthinum.** Notes, Critical and Explanatory, by W. M. DONNE, B.A., Trin. Coll., Cam. 1s. 6d.
9. **Terentii Andria et Heautontimorumenos.** With Notes, Critical and Explanatory, by the Rev. JAMES DAVIES, M.A. 1s. 6d.
10. **Terentii Adelphi, Hecyra, Phormio.** Edited, with Notes, Critical and Explanatory, by the Rev. JAMES DAVIES, M.A. 2s.
11. **Terentii Eunuchus, Comœdia.** Notes, by Rev. J. DAVIES, M.A. 1s. 6d.
12. **Ciceronis Oratio pro Sexto Roscio Amerino.** Edited, with an Introduction, Analysis, and Notes, Explanatory and Critical, by the Rev. JAMES DAVIES, M.A. 1s. 6d.
13. **Ciceronis Orationes in Catilinam, Verrem, et pro Archia.** With Introduction, Analysis, and Notes, Explanatory and Critical, by Rev. T. H. L. LEARY, D.C.L. formerly Scholar of Brasenose College, Oxford. 1s. 6d.
14. **Ciceronis Cato Major, Lælius, Brutus, sive de Senectute, de Amicitia, de Claris Oratoribus Dialogi.** With Notes by W. BROWNRIGG SMITH, M.A., F.R.G.S. 2s.
16. **Livy;** History of Rome. Notes by H. YOUNG and W. B. SMITH, M.A. Part 1. Books i., ii., 1s. 6d.
16*. ——— Part 2. Books iii., iv., v., 1s. 6d.
17. ——— Part 3. Books xxi., xxii., 1s. 6d.
19. **Latin Verse Selections,** from Catullus, Tibullus, Propertius, and Ovid. Notes by W. B. DONNE, M.A., Trinity College, Cambridge. 2s.
20. **Latin Prose Selections,** from Varro, Columella, Vitruvius, Seneca, Quintilian, Florus, Velleius Paterculus, Valerius Maximus Suetonius, Apuleius, &c. Notes by W. B. DONNE, M.A. 2s.
21. **Juvenalis Satiræ.** With Prolegomena and Notes by T. H. S. ESCOTT, B.A., Lecturer on Logic at King's College, London. 2s.

7, STATIONERS' HALL COURT, LUDGATE HILL, E.C.

GREEK.

14. **Greek Grammar,** in accordance with the Principles and Philological Researches of the most eminent Scholars of our own day. By HANS CLAUDE HAMILTON. 1s. 6d.

15,17. **Greek Lexicon.** Containing all the Words in General Use, with their Significations, Inflections, and Doubtful Quantities. By HENRY R. HAMILTON. Vol. 1. Greek-English, 2s. 6d.; Vol. 2. English-Greek, 2s. Or the Two Vols. in One, 4s. 6d.: cloth boards, 5s.

14,15, 17. **Greek Lexicon** (as above). Complete, with the GRAMMAR, in One Vol., cloth boards, 6s.

GREEK CLASSICS. With Explanatory Notes in English.

1. **Greek Delectus.** Containing Extracts from Classical Authors, with Genealogical Vocabularies and Explanatory Notes, by H. YOUNG. New Edition, with an improved and enlarged Supplementary Vocabulary, by JOHN HUTCHISON, M.A., of the High School, Glasgow. 1s. 6d.

2, 3. **Xenophon's Anabasis;** or, The Retreat of the Ten Thousand. Notes and a Geographical Register, by H. YOUNG. Part 1. Books i. to iii., 1s. Part 2. Books iv. to vii., 1s.

4. **Lucian's Select Dialogues.** The Text carefully revised, with Grammatical and Explanatory Notes, by H. YOUNG. 1s. 6d.

5-12. **Homer, The Works of.** According to the Text of BAEUMLEIN. With Notes, Critical and Explanatory, drawn from the best and latest Authorities, with Preliminary Observations and Appendices, by T. H. L. LEARY, M.A., D.C.L.

THE ILIAD: Part 1. Books i. to vi., 1s. 6d. | Part 3. Books xiii. to xviii., 1s. 6d.
Part 2. Books vii. to xii., 1s. 6d. | Part 4. Books xix. to xxiv., 1s. 6d.
THE ODYSSEY: Part 1. Books i. to vi., 1s. 6d | Part 3. Books xiii. to xviii., 1s. 6d.
Part 2. Books vii. to xii., 1s. 6d. | Part 4. Books xix. to xxiv., and Hymns, 2s.

13. **Plato's Dialogues:** The Apology of Socrates, the Crito, and the Phædo. From the Text of C. F. HERMANN. Edited with Notes, Critical and Explanatory, by the Rev. JAMES DAVIES, M.A. 2s.

14-17. **Herodotus, The History of,** chiefly after the Text of GAISFORD. With Preliminary Observations and Appendices, and Notes, Critical and Explanatory, by T. H. L. LEARY, M.A., D.C.L.
Part 1. Books i., ii. (The Clio and Euterpe), 2s.
Part 2. Books iii., iv. (The Thalia and Melpomene), 2s.
Part 3. Books v.-vii. (The Terpsichore, Erato, and Polymnia), 2s.
Part 4. Books viii., ix. (The Urania and Calliope) and Index, 1s. 6d.

18. **Sophocles:** Œdipus Tyrannus. Notes by H. YOUNG. 1s.

20. **Sophocles:** Antigone. From the Text of DINDORF. Notes, Critical and Explanatory, by the Rev. JOHN MILNER, B.A. 2s.

23. **Euripides:** Hecuba and Medea. Chiefly from the Text of DINDORF. With Notes, Critical and Explanatory, by W. BROWNRIGG SMITH, M.A., F.R.G.S. 1s. 6d.

26. **Euripides:** Alcestis. Chiefly from the Text of DINDORF. With Notes, Critical and Explanatory, by JOHN MILNER, B.A. 1s. 6d.

30. **Æschylus:** Prometheus Vinctus: The Prometheus Bound. From the Text of DINDORF. Edited, with English Notes, Critical and Explanatory, by the Rev. JAMES DAVIES, M.A. 1s.

32. **Æschylus:** Septem Contra Thebes: The Seven against Thebes. From the Text of DINDORF. Edited, with English Notes, Critical and Explanatory, by the Rev. JAMES DAVIES, M.A. 1s.

40. **Aristophanes:** Acharnians. Chiefly from the Text of C. H. WEISE. With Notes, by C. S. T. TOWNSHEND, M.A. 1s. 6d.

41. **Thucydides:** History of the Peloponnesian War. Notes by H. YOUNG. Book 1. 1s. 6d.

42. **Xenophon's Panegyric on Agesilaus.** Notes and Introduction by LL. F. W. JEWITT. 1s. 6d.

43. **Demosthenes.** The Oration on the Crown and the Philippics. With English Notes. By Rev. T. H. L. LEARY, D.C.L., formerly Scholar of Brasenose College, Oxford. 1s. 6d.

7, Stationers' Hall Court, London, E.C.
February, 1892.

A CATALOGUE OF BOOKS

INCLUDING NEW AND STANDARD WORKS IN

ENGINEERING: CIVIL, MECHANICAL, AND MARINE, MINING AND METALLURGY, ELECTRICITY AND ELECTRICAL ENGINEERING, ARCHITECTURE AND BUILDING, INDUSTRIAL AND DECORATIVE ARTS, SCIENCE, TRADE AGRICULTURE, GARDENING, LAND AND ESTATE MANAGEMENT, LAW, &c.

PUBLISHED BY

CROSBY LOCKWOOD & SON.

MECHANICAL ENGINEERING, etc.

New Pocket-Book for Mechanical Engineers.

THE MECHANICAL ENGINEER'S POCKET-BOOK OF TABLES, FORMULÆ, RULES AND DATA. A Handy Book of Reference for Daily Use in Engineering Practice. By D. KINNEAR CLARK, M.Inst.C.E., Author of "Railway Machinery," "Tramways," &c. &c. Small 8vo, nearly 700 pages. With Illustrations. Rounded edges, cloth limp, 7s. 6d.; or leather, gilt edges, 9s. [*Just published.*

New Manual for Practical Engineers.

THE PRACTICAL ENGINEER'S HAND-BOOK. Comprising a Treatise on Modern Engines and Boilers: Marine, Locomotive and Stationary. And containing a large collection of Rules and Practical Data relating to recent Practice in Designing and Constructing all kinds of Engines, Boilers, and other Engineering work. The whole constituting a comprehensive Key to the Board of Trade and other Examinations for Certificates of Competency in Modern Mechanical Engineering. By WALTER S. HUTTON, Civil and Mechanical Engineer, Author of "The Works' Manager's Handbook for Engineers," &c. With upwards of 370 Illustrations. Fourth Edition, Revised, with Additions. Medium 8vo, nearly 500 pp., price 18s. Strongly bound. [*Just published.*

☞ *This work is designed as a companion to the Author's* "WORKS' MANAGER'S HAND-BOOK." *It possesses many new and original features, and contains, like its predecessor, a quantity of matter not originally intended for publication, but collected by the author for his own use in the construction of a great variety of modern engineering work.*

⁎⁎⁎ OPINIONS OF THE PRESS.

"A thoroughly good practical handbook, which no engineer can go through without learning something that will be of service to him."—*Marine Engineer.*

"An excellent book of reference for engineers, and a valuable text-book for students of engineering."—*Scotsman.*

"This valuable manual embodies the results and experience of the leading authorities on mechanical engineering."—*Building News.*

"The author has collected together a surprising quantity of rules and practical data, and has shown much judgment in the selections he has made. . . . There is no doubt that this book is one of the most useful of its kind published, and will be a very popular compendium."—*Engineer.*

"A mass of information, set down in simple language, and in such a form that it can be easily referred to at any time. The matter is uniformly good and well chosen, and is greatly elucidated by the illustrations. The book will find its way on to most engineers' shelves, where it will rank as one of the most useful books of reference."—*Practical Engineer.*

"Should be found on the office shelf of all practical engineers."—*English Mechanic.*

B

Handbook for Works' Managers.

THE WORKS' MANAGER'S HANDBOOK OF MODERN RULES, TABLES, AND DATA. For Engineers, Millwrights, and Boiler Makers; Tool Makers, Machinists, and Metal Workers; Iron and Brass Founders, &c. By W. S. HUTTON, C.E., Author of "The Practical Engineer's Handbook." Fourth Edition, carefully Revised, and partly Re-written. In One handsome Volume, medium 8vo, 15s. strongly bound. [*Just published.*

☞ *The Author having compiled Rules and Data for his own use in a great variety of modern engineering work, and having found his notes extremely useful, decided to publish them—revised to date—believing that a practical work, suited to the* DAILY REQUIREMENTS OF MODERN ENGINEERS, *would be favourably received.*

In the Third Edition, the following among other additions have been made, viz.: Rules for the Proportions of Riveted Joints in Soft Steel Plates, the Results of Experiments by PROFESSOR KENNEDY *for the Institution of Mechanical Engineers—Rules for the Proportions of Turbines—Rules for the Strength of Hollow Shafts of Whitworth's Compressed Steel, &c.*

*** OPINIONS OF THE PRESS.

"The author treats every subject from the point of view of one who has collected workshop notes for application in workshop practice, rather than from the theoretical or literary aspect. The volume contains a great deal of that kind of information which is gained only by practical experience, and is seldom written in books."—*Engineer.*

"The volume is an exceedingly useful one, brimful with engineers' notes, memoranda, and rules, and well worthy of being on every mechanical engineer's bookshelf."—*Mechanical World.*

"The information is precisely that likely to be required in practice. . . . The work forms a desirable addition to the library not only of the works manager, but of anyone connected with general engineering."—*Mining Journal.*

"A formidable mass of facts and figures, readily accessible through an elaborate index . . . Such a volume will be found absolutely necessary as a book of reference in all sorts of 'works' connected with the metal trades."—*Ryland's Iron Trades Circular.*

"Brimful of useful information, stated in a concise form, Mr. Hutton's books have met a pressing want among engineers. The book must prove extremely useful to every practical man possessing a copy."—*Practical Engineer.*

Practical Treatise on Modern Steam-Boilers.

STEAM-BOILER CONSTRUCTION. A Practical Handbook for Engineers, Boiler-Makers, and Steam Users. Containing a large Collection of Rules and Data relating to the Design, Construction, and Working of Modern Stationary, Locomotive, and Marine Steam-Boilers. By WALTER S. HUTTON, C.E., Author of "The Works' Manager's Handbook," &c. With upwards of 300 Illustrations. Medium 8vo, 18s. cloth. [*Just published.*

"Every detail, both in boiler design and management, is clearly laid before the reader. The volume shows that boiler construction has been reduced to the condition of one of the most exact sciences; and such a book is of the utmost value to the *fin de siècle* Engineer and Works' Manager."—*Marine Engineer.*

"There has long been room for a modern handbook on steam boilers; there is not that room now, because Mr. Hutton has filled it. It is a thoroughly practical book for those who are occupied in the construction, design, selection, or use of boilers."—*Engineer.*

"The Modernised Templeton."

THE PRACTICAL MECHANIC'S WORKSHOP COMPANION. Comprising a great variety of the most useful Rules and Formulæ in Mechanical Science, with numerous Tables of Practical Data and Calculated Results for Facilitating Mechanical Operations. By WILLIAM TEMPLETON, Author of "The Engineer's Practical Assistant," &c. &c. Sixteenth Edition, Revised, Modernised, and considerably Enlarged by WALTER S. HUTTON, C.E., Author of "The Works' Manager's Handbook," "The Practical Engineer's Handbook," &c. Fcap. 8vo, nearly 500 pp., with Eight Plates and upwards of 250 Illustrative Diagrams, 6s., strongly bound for workshop or pocket wear and tear. [*Just published.*

*** OPINIONS OF THE PRESS.

"In its modernised form Hutton's 'Templeton' should have a wide sale, for it contains much valuable information which the mechanic will often find of use, and not a few tables and notes which he might look for in vain in other works. This modernised edition will be appreciated by all who have learned to value the original editions of 'Templeton.'"—*English Mechanic.*

"It has met with great success in the engineering workshop, as we can testify; and there are a great many men who, in a great measure, owe their rise in life to this little book."—*Building News.*

"This familiar text-book—well known to all mechanics and engineers—is of essential service to the every-day requirements of engineers, millwrights, and the various trades connected with engineering and building. The new modernised edition is worth its weight in gold."—*Building News.* (Second Notice.)

"This well-known and largely used book contains information, brought up to date, of the sort so useful to the foreman and draughtsman. So much fresh information has been introduced as to constitute it practically a new book. It will be largely used in the office and workshop."—*Mechanical World.*

MECHANICAL ENGINEERING, etc.

Stone-working Machinery.
STONE-WORKING MACHINERY, and the Rapid and Economical Conversion of Stone. With Hints on the Arrangement and Management of Stone Works. By M. POWIS BALE, M.I.M.E. With Illusts. Crown 8vo, 9s.

"Should be in the hands of every mason or student of stone-work."—*Colliery Guardian.*
"A capital handbook for all who manipulate stone for building or ornamental purposes."—*Machinery Market.*

Pump Construction and Management.
PUMPS AND PUMPING: A Handbook for Pump Users. Being Notes on Selection, Construction and Management. By M. POWIS BALE, M.I.M.E., Author of "Woodworking Machinery," &c. Crown 8vo, 2s. 6d.

"The matter is set forth as concisely as possible. In fact, condensation rather than diffuseness has been the author's aim throughout; yet he does not seem to have omitted anything likely to be of use."—*Journal of Gas Lighting.*

Milling Machinery, etc.
MILLING MACHINES AND PROCESSES: A Practical Treatise on Shaping Metals by Rotary Cutters, including Information on Making and Grinding the Cutters. By PAUL N. HASLUCK, Author of "Lathe-work." With upwards of 300 Engravings. Large crown 8vo, 12s. 6d. cloth.
[*Just published.*

Turning.
LATHE-WORK: A Practical Treatise on the Tools, Appliances, and Processes employed in the Art of Turning. By PAUL N. HASLUCK. Fourth Edition, Revised and Enlarged. Cr. 8vo, 5s. cloth.

"Written by a man who knows, not only how work ought to be done, but who also knows how to do it, and how to convey his knowledge to others. To all turners this book would be valuable."—*Engineering.*
"We can safely recommend the work to young engineers. To the amateur it will simply be invaluable. To the student it will convey a great deal of useful information."—*Engineer.*

Screw-Cutting.
SCREW THREADS: And Methods of Producing Them. With Numerous Tables, and complete directions for using Screw-Cutting Lathes. By PAUL N. HASLUCK, Author of "Lathe-Work," &c. With Fifty Illustrations. Third Edition, Enlarged. Waistcoat-pocket size, 1s. 6d. cloth.

"Full of useful information, hints and practical criticism. Taps, dies and screwing-tools generally are illustrated and their action described."—*Mechanical World.*
"It is a complete compendium of all the details of the screw-cutting lathe; in fact a *multum-in-parvo* on all the subjects it treats upon."—*Carpenter and Builder.*

Smith's Tables for Mechanics, etc.
TABLES, MEMORANDA, AND CALCULATED RESULTS, FOR MECHANICS, ENGINEERS, ARCHITECTS, BUILDERS, etc. Selected and Arranged by FRANCIS SMITH. Fifth Edition, thoroughly Revised and Enlarged, with a New Section of ELECTRICAL TABLES, FORMULÆ, and MEMORANDA. Waistcoat-pocket size, 1s. 6d. limp leather. [*Just published.*

"It would, perhaps, be as difficult to make a small pocket-book selection of notes and formulæ to suit ALL engineers as it would be to make a universal medicine; but Mr. Smith's waistcoat-pocket collection may be looked upon as a successful attempt."—*Engineer.*
"The best example we have ever seen of 250 pages of useful matter packed into the dimensions of a card-case."—*Building News.* "A veritable pocket treasury of knowledge."—*Iron.*

Engineer's and Machinist's Assistant.
THE ENGINEER'S, MILLWRIGHT'S, and MACHINIST'S PRACTICAL ASSISTANT. A collection of Useful Tables, Rules and Data. By WILLIAM TEMPLETON. 7th Edition, with Additions. 18mo, 2s. 6d. cloth.

"Occupies a foremost place among books of this kind. A more suitable present to an apprentice to any of the mechanical trades could not possibly be made."—*Building News.*
"A deservedly popular work, it should be in the 'drawer' of every mechanic."—*English Mechanic.*

Iron and Steel.
"IRON AND STEEL": A Work for the Forge, Foundry, Factory, and Office. Containing ready, useful, and trustworthy Information for Ironmasters; Managers of Bar, Rail, Plate, and Sheet Rolling Mills; Iron and Metal Founders; Iron Ship and Bridge Builders; Mechanical, Mining, and Consulting Engineers; Contractors, Builders, &c. By CHARLES HOARE. Eighth Edition, Revised and considerably Enlarged. 32mo, 6s. leather.

"One of the best of the pocket books."—*English Mechanic.*
"We cordially recommend this book to those engaged in considering the details of all kinds of iron and steel works."—*Naval Science.*

Engineering Construction.

PATTERN-MAKING: A Practical Treatise, embracing the Main Types of Engineering Construction, and including Gearing, both Hand and Machine made, Engine Work, Sheaves and Pulleys, Pipes and Columns, Screws, Machine Parts, Pumps and Cocks, the Moulding of Patterns in Loam and Greensand, &c., together with the methods of Estimating the weight of Castings; to which is added an Appendix of Tables for Workshop Reference. By a FOREMAN PATTERN MAKER. With upwards of Three Hundred and Seventy Illustrations. Crown 8vo, 7s. 6d. cloth.

"A well-written technical guide, evidently written by a man who understands and has practised what he has written about. . . . We cordially recommend it to engineering students, young journeymen, and others desirous of being initiated into the mysteries of pattern-making."—*Builder.*

"We can confidently recommend this comprehensive treatise."—*Building News.*

"Likely to prove a welcome guide to many workmen, especially to draughtsmen who have lacked a training in the shops, pupils pursuing their practical studies in our factories, and to employers and managers in engineering works."—*Hardware Trade Journal.*

"More than 370 illustrations help to explain the text, which is, however, always clear and explicit, thus rendering the work an excellent *vade mecum* for the apprentice who desires to become master of his trade."—*English Mechanic.*

Dictionary of Mechanical Engineering Terms.

LOCKWOOD'S DICTIONARY OF TERMS USED IN THE PRACTICE OF MECHANICAL ENGINEERING, embracing those current in the Drawing Office, Pattern Shop, Foundry, Fitting, Turning, Smith's and Boiler Shops, &c. &c. Comprising upwards of 6,000 Definitions. Edited by A FOREMAN PATTERN-MAKER, Author of "Pattern Making." Crown 8vo, 7s. 6d. cloth.

"Just the sort of handy dictionary required by the various trades engaged in mechanical engineering. The practical engineering pupil will find the book of great value in his studies, and every foreman engineer and mechanic should have a copy."—*Building News.*

"After a careful examination of the book, and trying all manner of words, we think that the engineer will here find all he is likely to require. It will be largely used."—*Practical Engineer.*

"One of the most useful books which can be presented to a mechanic or student."—*English Mechanic.*

"Not merely a dictionary, but, to a certain extent, also a most valuable guide. It strikes us as a happy idea to combine with a definition of the phrase useful information on the subject of which it treats."—*Machinery Market.*

"No word having connection with any branch of constructive engineering seems to be omitted. No more comprehensive work has been, so far, issued. —*Knowledge.*

"We strongly commend this useful and reliable adviser to our friends in the workshop, and to students everywhere."—*Colliery Guardian.*

Steam Boilers.

A TREATISE ON STEAM BOILERS: Their Strength, Construction, *and Economical Working.* By ROBERT WILSON, C.E. Fifth Edition. 12mo, 6s. cloth.

"The best treatise that has ever been published on steam boilers."—*Engineer.*

"The author shows himself perfect master of his subject, and we heartily recommend all employing steam power to possess themselves of the work."—*Ryland's Iron Trade Circular.*

Boiler Chimneys.

BOILER AND FACTORY CHIMNEYS; *Their Draught-Power and Stability.* With a Chapter on *Lightning Conductors.* By ROBERT WILSON, A.I.C.E., Author of "A Treatise on Steam Boilers," &c. Second Edition. Crown 8vo, 3s. 6d. cloth.

"Full of useful information, definite in statement, and thoroughly practical in treatment. —*The Local Government Chronicle.*

"A valuable contribution to the iterature of scientific building."—*The Builder.*

Boiler Making.

THE BOILER-MAKER'S READY RECKONER & ASSIST-ANT. With Examples of Practical Geometry and Templating, for the Use of Platers, Smiths and Riveters. By JOHN COURTNEY, Edited by D. K. CLARK, M.I.C.E. Third Edition, 480 pp., with 140 Illusts. Fcap. 8vo, 7s. half-bound.

"No workman or apprentice should be without this book."—*Iron Trade Circular.*

"Boiler-makers will readily recognise the value of this volume. . . . The tables are clearly printed, and so arranged that they can be referred to with the greatest facility, so that it cannot be doubted that they will be generally appreciated and much used."—*Mining Journal.*

Warming.

HEATING BY HOT WATER; with Information and Suggestions on the best Methods of Heating Public, Private and Horticultural Buildings. By WALTER JONES. With Illustrations, crown 8vo, 2s. cloth.

"We confidently recommend all interested in heating by hot water to secure a copy of thi valuable little treatise."—*The Plumber and Decorator.*

Steam Engine.

TEXT-BOOK ON THE STEAM ENGINE. With a Supplement on Gas Engines, and PART II. ON HEAT ENGINES. By T. M. GOODEVE, M.A., Barrister-at-Law, Professor of Mechanics at the Normal School of Science and the Royal School of Mines; Author of "The Principles of Mechanics," "The Elements of Mechanism," &c. Eleventh Edition, Enlarged. With numerous Illustrations. Crown 8vo, 6s. cloth.

"Professor Goodeve has given us a treatise on the steam engine which will bear comparison with anything written by Huxley or Maxwell, and we can award it no higher praise."—*Engineer.*

" Mr. Goodeve's text-book is a work of which every young engineer should possess himself.' —*Mining Journal.*

Gas Engines.

ON GAS-ENGINES. Being a Reprint, with some Additions, of the Supplement to the *Text-book on the Steam Engine*, by T. M. GOODEVE, M.A. Crown 8vo, 2s. 6d. cloth.

" Like all Mr. Goodeve's writings, the present is no exception in point of general excellence. It is a valuable little volume."—*Mechanical World.*

Steam.

THE SAFE USE OF STEAM. Containing Rules for Unprofessional Steam-users. By an ENGINEER. Sixth Edition. Sewed, 6d.

"If steam-users would but learn this little book by heart boiler explosions would become sensations by their rarity."—*English Mechanic.*

Reference Book for Mechanical Engineers.

THE MECHANICAL ENGINEER'S REFERENCE BOOK, for Machine and Boiler Construction. In Two Parts. Part I. GENERAL ENGINEERING DATA. Part II. BOILER CONSTRUCTION. With 51 Plates and numerous Illustrations. By NELSON FOLEY, M.I.N.A. Folio, £5 5s. half-bound.
[*Just published.*

Coal and Speed Tables.

A POCKET BOOK OF COAL AND SPEED TABLES, for Engineers and Steam-users. By NELSON FOLEY, Author of "Boiler Construction." Pocket-size, 3s. 6d. cloth; 4s. leather.

"These tables are designed to meet the requirements of every-day use; and may be commended to engineers and users of steam."—*Iron.*

"This pocket-book well merits the attention of the practical engineer. Mr. Foley has compiled a very useful set of tables, the information contained in which is frequently required by engineers, coal consumers and users of steam."—*Iron and Coal Trades Review.*

Fire Engineering.

FIRES, FIRE-ENGINES, AND FIRE-BRIGADES. With a History of Fire-Engines, their Construction, Use, and Management; Remarks on Fire-Proof Buildings, and the Preservation of Life from Fire; Foreign Fire Systems, &c. By C. F. T. YOUNG, C.E. With numerous Illustrations, 544 pp., demy 8vo, £1 4s. cloth.

"To such of our readers as are interested in the subject of fires and fire apparatus, we can most heartily commend this book."—*Engineering.*

"It displays much evidence of careful research; and Mr. Young has put his facts neatly together. It is evident enough that his acquaintance with the practical details of the construction of steam fire engines is accurate and full."—*Engineer.*

Estimating for Engineering Work, &c.

ENGINEERING ESTIMATES, COSTS AND ACCOUNTS: A Guide to Commercial Engineering. With numerous Examples of Estimates and Costs of Millwright Work, Miscellaneous Productions, Steam Engines and Steam Boilers; and a Section on the Preparation of Costs Accounts. By A GENERAL MANAGER. Demy 8vo, 12s. cloth.

"This is an excellent and very useful book, covering subject-matter in constant requisition in every factory and workshop. . . . The book is invaluable, not only to the young engineer, but also to the estimate department of every works."—*Builder.*

"We accord the work unqualified praise. The information is given in a plain, straightforward manner, and bears throughout evidence of the intimate practical acquaintance of the author with every phrase of commercial engineering."—*Mechanical World.*

Elementary Mechanics.

CONDENSED MECHANICS. A Selection of Formulæ, Rules, Tables, and Data for the Use of Engineering Students, Science Classes, &c. In Accordance with the Requirements of the Science and Art Department By W. G. CRAWFORD HUGHES, A.M.I.C.E. Crown 8vo, 2s. 6d. cloth.
[*Just published.*

THE POPULAR WORKS OF MICHAEL REYNOLDS
("THE ENGINE DRIVER'S FRIEND").

Locomotive-Engine Driving.

LOCOMOTIVE-ENGINE DRIVING: *A Practical Manual for Engineers in charge of Locomotive Engines.* By MICHAEL REYNOLDS, Member of the Society of Engineers, formerly Locomotive Inspector L. B. and S. C. R. Eighth Edition. Including a KEY TO THE LOCOMOTIVE ENGINE. With Illustrations and Portrait of Author. Crown 8vo, 4s. 6d. cloth.

"Mr. Reynolds has supplied a want, and has supplied it well. We can confidently recommend the book, not only to the practical driver, but to everyone who takes an interest in the performance of locomotive engines."—*The Engineer.*

"Mr. Reynolds has opened a new chapter in the literature of the day. This admirable practical treatise, of the practical utility of which we have to speak in terms of warm commendation."—*Athenæum.*

"Evidently the work of one who knows his subject thoroughly."—*Railway Service Gazette.*

"Were the cautions and rules given in the book to become part of the every-day working of our engine-drivers, we might have fewer distressing accidents to deplore."—*Scotsman.*

Stationary Engine Driving.

STATIONARY ENGINE DRIVING: *A Practical Manual for Engineers in charge of Stationary Engines.* By MICHAEL REYNOLDS. Fourth Edition, Enlarged. With Plates and Woodcuts. Crown 8vo, 4s. 6d. cloth.

"The author is thoroughly acquainted with his subjects, and his advice on the various points treated is clear and practical. . . . He has produced a manual which is an exceedingly useful one for the class for whom it is specially intended."—*Engineering.*

"Our author leaves no stone unturned. He is determined that his readers shall not only know something about the stationary engine, but all about it."—*Engineer.*

"An engineman who has mastered the contents of Mr. Reynolds's book will require but little actual experience with boilers and engines before he can be trusted to look after them."—*English Mechanic.*

The Engineer, Fireman, and Engine-Boy.

THE MODEL LOCOMOTIVE ENGINEER, FIREMAN, and ENGINE-BOY. Comprising a Historical Notice of the Pioneer Locomotive Engines and their Inventors. By MICHAEL REYNOLDS. With numerous Illustrations and a fine Portrait of George Stephenson. Crown 8vo, 4s. 6d. cloth.

"From the technical knowledge of the author it will appeal to the railway man of to-day more forcibly than anything written by Dr. Smiles. . . . The volume contains information of a technical kind, and facts that every driver should be familiar with."—*English Mechanic.*

"We should be glad to see this book in the possession of everyone in the kingdom who has ever laid, or is to lay, hands on a locomotive engine."—*Iron.*

Continuous Railway Brakes.

CONTINUOUS RAILWAY BRAKES: *A Practical Treatise on the several Systems in Use in the United Kingdom; their Construction and Performance.* With copious Illustrations and numerous Tables. By MICHAEL REYNOLDS. Large crown 8vo, 9s. cloth.

"A popular explanation of the different brakes. It will be of great assistance in forming public opinion, and will be studied with benefit by those who take an interest in the brake."—*English Mechanic.*

"Written with sufficient technical detail to enable the principle and relative connection of the various parts of each particular brake to be readily grasped."—*Mechanical World.*

Engine-Driving Life.

ENGINE-DRIVING LIFE: *Stirring Adventures and Incidents in the Lives of Locomotive-Engine Drivers.* By MICHAEL REYNOLDS. Second Edition, with Additional Chapters. Crown 8vo, 2s. cloth.

"From first to last perfectly fascinating. Wilkie Collins's most thrilling conceptions are thrown into the shade by true incidents, endless in their variety, related in every page."—*North British Mail.*

"Anyone who wishes to get a real insight into railway life cannot do better than read 'Engine-Driving Life' for himself; and if he once take it up he will find that the author's enthusiasm and real love of the engine-driving profession will carry him on till he has read every page."—*Saturday Review.*

Pocket Companion for Enginemen.

THE ENGINEMAN'S POCKET COMPANION AND PRACTICAL EDUCATOR FOR ENGINEMEN, BOILER ATTENDANTS, AND MECHANICS. By MICHAEL REYNOLDS. With Forty-five Illustrations and numerous Diagrams. Second Edition, Revised. Royal 18mo, 3s. 6d., strongly bound for pocket wear.

"This admirable work is well suited to accomplish its object, being the honest workmanship of a competent engineer."—*Glasgow Herald.*

"A most meritorious work, giving in a succinct and practical form all the information an engineminder desirous of mastering the scientific principles of his daily calling would require."—*Miller.*

"A boon to those who are striving to become efficient mechanics."—*Daily Chronicle.*

French-English Glossary for Engineers, etc.
A POCKET GLOSSARY of TECHNICAL TERMS: ENGLISH-FRENCH, FRENCH-ENGLISH; with Tables suitable for the Architectural, Engineering, Manufacturing and Nautical Professions. By JOHN JAMES FLETCHER, Engineer and Surveyor. 200 pp. Waistcoat-pocket size, 1s. 6d., limp leather.

"It ought certainly to be in the waistcoat-pocket of every professional man."—*Iron.*

"It is a very great advantage for readers and correspondents in France and England to have so large a number of the words relating to engineering and manufacturers collected in a liliputian volume. The little book will be useful both to students and travellers.'—*Architect.*

"The glossary of terms is very complete, and many of the tables are new and well arranged. We cordially commend the book."—*Mechanical World.*

Portable Engines.
THE PORTABLE ENGINE; ITS CONSTRUCTION AND MANAGEMENT. A Practical Manual for Owners and Users of Steam Engines generally. By WILLIAM DYSON WANSBROUGH. With 90 Illustrations. Crown 8vo, 3s. 6d. cloth.

"This is a work of value to those who use steam machinery. . . . Should be read by every-one who has a steam engine, on a farm or elsewhere."—*Mark Lane Express.*

"We cordially commend this work to buyers and owners of steam engines, and to those who have to do with their construction or use."—*Timber Trades Journal.*

"Such a general knowledge of the steam engine as Mr. Wansbrough furnishes to the reader should be acquired by all intelligent owners and others who use the steam engine."—*Building News.*

"An excellent text-book of this useful form of engine, which describes with all necessary minuteness the details of the various devices. . . 'The Hints to Purchasers contain a good deal of commonsense and practical wisdom.'"—*English Mechanic.*

CIVIL ENGINEERING, SURVEYING, etc.

MR. HUMBER'S IMPORTANT ENGINEERING BOOKS.

The Water Supply of Cities and Towns.
A COMPREHENSIVE TREATISE on the WATER-SUPPLY OF CITIES AND TOWNS. By WILLIAM HUMBER, A-M.Inst.C.E., and M. Inst. M.E., Author of "Cast and Wrought Iron Bridge Construction," &c. &c. Illustrated with 50 Double Plates, 1 Single Plate, Coloured Frontispiece, and upwards of 250 Woodcuts, and containing 400 pages of Text. Imp. 4to, £6 6s. elegantly and substantially half-bound in morocco.

List of Contents.

I. Historical Sketch of some of the means that have been adopted for the Supply of Water to Cities and Towns.—II. Water and the Foreign Matter usually associated with it.—III. Rainfall and Evaporation.—IV. Springs and the water-bearing formations of various districts.—V. Measurement and Estimation of the flow of Water.—VI. On the Selection of the Source of Supply.—VII. Wells.—VIII. Reservoirs.—IX. The Purification of Water.—X. Pumps.—XI. Pumping Machinery.—XII. Conduits.—XIII. Distribution of Water.—XIV. Meters, Service Pipes, and House Fittings.—XV. The Law and Economy of Water Works. XVI. Constant and Intermittent Supply.—XVII. Description of Plates.—Appendices, giving Tables of Rates of Supply, Velocities, &c. &c., together with Specifications of several Works illustrated, among which will be found: Aberdeen, Bideford, Canterbury, Dundee, Halifax, Lambeth, Rotherham, Dublin, and others.

"The most systematic and valuable work upon water supply hitherto produced in English, or in any other language. . . . Mr. Humber's work is characterised almost throughout by an exhaustiveness much more distinctive of French and German than of English technical treatises."—*Engineer.*

"We can congratulate Mr. Humber on having been able to give so large an amount of information on a subject so important as the water supply of cities and towns. The plates, fifty in number, are mostly drawings of executed works, and alone would have commanded the attention of every engineer whose practice may lie in this branch of the profession."—*Builder.*

Cast and Wrought Iron Bridge Construction.
A COMPLETE AND PRACTICAL TREATISE ON CAST AND WROUGHT IRON BRIDGE CONSTRUCTION, including Iron Foundations. In Three Parts—Theoretical, Practical, and Descriptive. By WILLIAM HUMBER, A.M.Inst.C.E., and M.Inst.M.E. Third Edition, Revised and much improved, with 115 Double Plates (20 of which now first appear in this edition), and numerous Additions to the Text. In Two Vols., imp. 4to, £6 16s. 6d. half-bound in morocco.

"A very valuable contribution to the standard literature of civil engineering. In addition to elevations, plans and sections, large scale details are given which very much enhance the instructive worth of those illustrations."—*Civil Engineer and Architect's Journal.*

"Mr. Humber's stately volumes, lately issued—in which the most important bridges erected during the last five years, under the direction of the late Mr. Brunel, Sir W. Cubitt, Mr. Hawkshaw, Mr. Page, Mr. Fowler, Mr. Hemans, and others among our most eminent engineers, are drawn and specified in great detail."—*Engineer.*

MR. HUMBER'S GREAT WORK ON MODERN ENGINEERING.

Complete in Four Volumes, imperial 4to, price £12 12s., half-morocco. Each Volume sold separately as follows:—

A RECORD OF THE PROGRESS OF MODERN ENGINEERING. FIRST SERIES.
Comprising Civil, Mechanical, Marine, Hydraulic, Railway, Bridge, and other Engineering Works, &c. By WILLIAM HUMBER, A-M.Inst.C.E., &c. Imp. 4to, with 36 Double Plates, drawn to a large scale, Photographic Portrait of John Hawkshaw, C.E., F.R.S., &c., and copious descriptive Letterpress, Specifications, &c., £3 3s. half-morocco.

List of the Plates and Diagrams.

Victoria Station and Roof, L. B. & S. C. R. (8 plates); Southport Pier (2 plates); Victoria Station and Roof, L. C. & D. and G. W. R. (6 plates); Roof of Cremorne Music Hall; Bridge over G. N. Railway; Roof of Station, Dutch Rhenish Rail (2 plates); Bridge over the Thames, West London Extension Railway (5 plates); Armour Plates: Suspension Bridge, Thames (4 plates); The Allen Engine; Suspension Bridge, Avon (3 plates); Underground Railway (3 plates).

"Handsomely lithographed and printed. It will find favour with many who desire to preserve in a permanent form copies of the plans and specifications prepared for the guidance of the contractors for many important engineering works."—*Engineer.*

HUMBER'S RECORD OF MODERN ENGINEERING. SECOND SERIES.
Imp. 4to, with 36 Double Plates, Photographic Portrait of Robert Stephenson, C.E., M.P., F.R.S., &c., and copious descriptive Letterpress, Specifications, &c., £3 3s. half-morocco.

List of the Plates and Diagrams.

Birkenhead Docks, Low Water Basin (15 plates); Charing Cross Station Roof, C. C. Railway (3 plates); Digswell Viaduct, Great Northern Railway; Robbery Wood Viaduct, Great Northern Railway; Iron Permanent Way; Clydach Viaduct, Merthyr, Tredegar, and Abergavenny Railway; Ebbw Viaduct, Merthyr, Tredegar, and Abergavenny Railway; College Wood Viaduct, Cornwall Railway; Dublin Winter Palace Roof (3 plates); Bridge over the Thames, L. C. & D. Railway (6 plates); Albert Harbour, Greenock (4 plates).

"Mr. Humber has done the profession good and true service, by the fine collection of examples he has here brought before the profession and the public."—*Practical Mechanic's Journal.*

HUMBER'S RECORD OF MODERN ENGINEERING. THIRD SERIES.
Imp. 4to, with 40 Double Plates, Photographic Portrait of J. R. M'Clean, late Pres. Inst. C.E., and copious descriptive Letterpress, Specifications, &c., £3 3s. half-morocco.

List of the Plates and Diagrams.

MAIN DRAINAGE, METROPOLIS.—*North Side.*—Map showing Interception of Sewers; Middle Level Sewer (2 plates); Outfall Sewer, Bridge over River Lea (3 plates); Outfall Sewer, Bridge over Marsh Lane, North Woolwich Railway, and Bow and Barking Railway Junction; Outfall Sewer, Bridge over Bow and Barking Railway (3 plates); Outfall Sewer, Bridge over East London Waterworks' Feeder (2 plates); Outfall Sewer, Reservoir (2 plates); Outfall Sewer, Tumbling Bay and Outlet; Outfall Sewer, Penstocks. *South Side.*—Outfall Sewer, Bermondsey Branch (2 plates); Outfall Sewer, Reservoir and Outlet (4 plates); Outfall Sewer, Filth Hoist; Sections of Sewers (North and South Sides). THAMES EMBANKMENT.—Section of River Wall; Steamboat Pier, Westminster (2 plates); Landing Stairs between Charing Cross and Waterloo Bridges; York Gate (2 plates); Overflow and Outlet at Savoy Street Sewer (3 plates); Steamboat Pier, Waterloo Bridge (3 plates); Junction of Sewers, Plans and Sections; Gullies, Plans and Sections; Rolling Stock; Granite and Iron Forts.

"The drawings have a constantly increasing value, and whoever desires to possess clear representations of the two great works carried out by our Metropolitan Board will obtain Mr. Humber's volume."—*Engineer.*

HUMBER'S RECORD OF MODERN ENGINEERING. FOURTH SERIES.
Imp. 4to, with 36 Double Plates, Photographic Portrait of John Fowler, late Pres. Inst. C.E., and copious descriptive Letterpress, Specifications, &c., £3 3s. half-morocco.

List of the Plates and Diagrams.

Abbey Mills Pumping Station, Main Drainage, Metropolis (4 plates); Barrow Docks (5 plates); Manquis Viaduct, Santiago and Valparaiso Railway (2 plates); Adam's Locomotive, St. Helen's Canal Railway (2 plates); Cannon Street Station Roof, Charing Cross Railway (3 plates); Road Bridge over the River Moka (2 plates); Telegraphic Apparatus for Mesopotamia; Viaduct over the River Wye, Midland Railway (3 plates); St. Germans Viaduct, Cornwall Railway (2 plates); Wrought-Iron Cylinder for Diving Bell; Millwall Docks (6 plates); Milroy's Patent Excavator; Metropolitan District Railway (6 plates); Harbours, Ports, and Breakwaters (3 plates).

"We gladly welcome another year's issue of this valuable publication from the able pen of Mr. Humber. The accuracy and general excellence of this work are well known, while its usefulness in giving the measurements and details of some of the latest examples of engineering, as carried out by the most eminent men in the profession, cannot be too highly prized."—*Artisan.*

MR. HUMBER'S ENGINEERING BOOKS—continued.

Strains, Calculation of.
A HANDY BOOK FOR THE CALCULATION OF STRAINS IN GIRDERS AND SIMILAR STRUCTURES, AND THEIR STRENGTH. Consisting of Formulæ and Corresponding Diagrams, with numerous details for Practical Application, &c. By WILLIAM HUMBER, A-M.Inst.C.E., &c. Fifth Edition. Crown 8vo, nearly 100 Woodcuts and 3 Plates, 7s. 6d. cloth
"The formulæ are neatly expressed, and the diagrams good."—*Athenæum.*
"We heartily commend this really *handy* book to our engineer and architect readers."—*English Mechanic.*

Barlow's Strength of Materials, enlarged by Humber
A TREATISE ON THE STRENGTH OF MATERIALS; with Rules for Application in Architecture, the Construction of Suspension Bridges, Railways, &c. By PETER BARLOW, F.R.S. A New Edition, revised by his Sons, P. W. BARLOW, F.R.S., and W. H. BARLOW, F.R.S.; to which are added, Experiments by HODGKINSON, FAIRBAIRN, and KIRKALDY; and Formulæ for Calculating Girders, &c. Arranged and Edited by W. HUMBER, A-M.Inst.C.E. Demy 8vo, 400 pp., with 19 large Plates and numerous Woodcuts, 18s. cloth.
"Valuable alike to the student, tyro, and the experienced practitioner, it will always rank in future, as it has hitherto done, as the standard treatise on that particular subject."—*Engineer.*
"There is no greater authority than Barlow."—*Building News.*
"As a scientific work of the first class, it deserves a foremost place on the bookshelves of every civil engineer and practical mechanic."—*English Mechanic.*

Trigonometrical Surveying.
AN OUTLINE OF THE METHOD OF CONDUCTING A TRIGONOMETRICAL SURVEY, *for the Formation of Geographical and Topographical Maps and Plans, Military Reconnaissance, Levelling, &c.,* with Useful Problems, Formulæ, and Tables. By Lieut.-General FROME, R.E. Fourth Edition, Revised and partly Re-written by Major General Sir CHARLES WARREN, G.C.M.G., R.E. With 19 Plates and 115 Woodcuts, royal 8vo, 16s. cloth.
"The simple fact that a fourth edition has been called for is the best testimony to its merits. No words of praise from us can strengthen the position so well and so steadily maintained by this work. Sir Charles Warren has revised the entire work, and made such additions as were necessary to bring every portion of the contents up to the present date."—*Broad Arrow.*

Field Fortification.
A TREATISE ON FIELD FORTIFICATION, THE ATTACK OF FORTRESSES, MILITARY MINING, AND RECONNOITRING. By Colonel I. S. MACAULAY, late Professor of Fortification in the R.M.A., Woolwich. Sixth Edition, crown 8vo, cloth, with separate Atlas of 12 Plates, 12s.

Oblique Bridges.
A PRACTICAL AND THEORETICAL ESSAY ON OBLIQUE BRIDGES. With 13 large Plates. By the late GEORGE WATSON BUCK, M.I.C.E. Third Edition, revised by his Son, J. H. WATSON BUCK, M.I.C.E.; and with the addition of Description to Diagrams for Facilitating the Construction of Oblique Bridges, by W. H. BARLOW, M.I.C.E. Royal 8vo, 12s, cloth.
"The standard text-book for all engineers regarding skew arches is Mr. Buck's treatise, and would be impossible to consult a better."—*Engineer.*
"Mr. Buck's treatise is recognised as a standard text-book, and his treatment has divested the subject of many of the intricacies supposed to belong to it. As a guide to the engineer and architect, on a confessedly difficult subject, Mr. Buck's work is unsurpassed."—*Building News.*

Water Storage, Conveyance and Utilisation.
WATER ENGINEERING: A Practical Treatise on the Measurement, Storage, Conveyance and Utilisation of Water for the Supply of Towns, for Mill Power, and for other Purposes. By CHARLES SLAGG, Water and Drainage Engineer, A.M.Inst.C.E., Author of "Sanitary Work in the Smaller Towns, and in Villages," &c. With numerous Illusts. Cr. 8vo, 7s. 6d. cloth.
"As a small practical treatise on the water supply of towns, and on some applications of water-power, the work is in many respects excellent."—*Engineering.*
"The author has collated the results deduced from the experiments of the most eminent authorities, and has presented them in a compact and practical form, accompanied by very clear and detailed explanations. . . . The application of water as a motive power is treated very carefully and exhaustively."—*Builder.*
"For anyone who desires to begin the study of hydraulics with a consideration of the practical applications of the science there is no better guide."—*Architect.*

Statics, Graphic and Analytic.

GRAPHIC AND ANALYTIC STATICS, in their Practical Application to the Treatment of Stresses in Roofs, Solid Girders, Lattice, Bowstring and Suspension Bridges, Braced Iron Arches and Piers, and other Frameworks. By R. HUDSON GRAHAM, C.E. Containing Diagrams and Plates to Scale. With numerous Examples, many taken from existing Structures. Specially arranged for Class-work in Colleges and Universities. Second Edition, Revised and Enlarged. 8vo, 16s. cloth.

"Mr. Graham's book will find a place wherever graphic and analytic statics are used or studied."—*Engineer.*

"The work is excellent from a practical point of view, and has evidently been prepared with much care. The directions for working are ample, and are illustrated by an abundance of well-selected examples. It is an excellent text-book for the practical draughtsman."—*Athenæum.*

Student's Text-Book on Surveying.

PRACTICAL SURVEYING: A Text-Book for Students preparing for Examination or for Survey-work in the Colonies. By GEORGE W. USILL, A.M.I.C.E., Author of "The Statistics of the Water Supply of Great Britain." With Four Lithographic Plates and upwards of 330 Illustrations. Second Edition, Revised. Crown 8vo, 7s. 6d. cloth.

"The best forms of instruments are described as to their construction, uses and modes of employment, and there are innumerable hints on work and equipment such as the author, in his experience as surveyor, draughtsman and teacher, has found necessary, and which the student in his inexperience will find most serviceable."—*Engineer.*

"The latest treatise in the English language on surveying, and we have no hesitation in saying that the student will find it a better guide than any of its predecessors . . . Deserves to be recognised as the first book which should be put in the hands of a pupil of Civil Engineering, and every gentleman of education who sets out for the Colonies would find it well to have a copy."—*Architect.*

"A very useful, practical handbook on field practice. Clear, accurate and not too condensed."—*Journal of Education.*

Survey Practice.

AID TO SURVEY PRACTICE, for Reference in Surveying, Levelling, and Setting-out; and in Route Surveys of Travellers by Land and Sea. With Tables, Illustrations, and Records. By LOWIS D'A. JACKSON, A.M.I.C.E., Author of "Hydraulic Manual," "Modern Metrology," &c. Second Edition, Enlarged. Large crown 8vo, 12s. 6d. cloth.

"Mr. Jackson has produced a valuable vade-mecum for the surveyor. We can recommend this book as containing an admirable supplement to the teaching of the accomplished surveyor."—*Athenæum.*

"As a text-book we should advise all surveyors to place it in their libraries, and study well the matured instructions afforded in its pages."—*Colliery Guardian.*

"The author brings to his work a fortunate union of theory and practical experience which, aided by a clear and lucid style of writing, renders the book a very useful one."—*Builder.*

Surveying, Land and Marine.

LAND AND MARINE SURVEYING, in Reference to the Preparation of Plans for Roads and Railways; Canals, Rivers, Towns' Water Supplies; Docks and Harbours. With Description and Use of Surveying Instruments. By W. D. HASKOLL, C.E., Author of "Bridge and Viaduct Construction," &c. Second Edition, Revised, with Additions. Large cr. 8vo, 9s. cl.

"This book must prove of great value to the student. We have no hesitation in recommending it, feeling assured that it will more than repay a careful study."—*Mechanical World.*

"A most useful and well arranged book for the aid of a student. We can strongly recommend it as a carefully-written and valuable text-book. It enjoys a well-deserved repute among surveyors."—*Builder.*

"This volume cannot fail to prove of the utmost practical utility. It may be safely recommended to all students who aspire to become clean and expert surveyors."—*Mining Journal.*

Tunnelling.

PRACTICAL TUNNELLING. Explaining in detail the Setting-out of the works, Shaft-sinking and Heading-driving, Ranging the Lines and Levelling underground, Sub-Excavating, Timbering, and the Construction of the Brickwork of Tunnels, with the amount of Labour required for, and the Cost of, the various portions of the work. By FREDERICK W. SIMMS, F.G.S., M.Inst.C.E. Third Edition, Revised and Extended by D. KINNEAR CLARK, M.Inst.C.E. Imperial 8vo, with 21 Folding Plates and numerous Wood Engravings, 30s. cloth.

"The estimation in which Mr. Simms's book on tunnelling has been held for over thirty years cannot be more truly expressed than in the words of the late Prof. Rankine:—'The best source of information on the subject of tunnels is Mr. F. W. Simms's work on Practical Tunnelling.'"—*Architect.*

"It has been regarded from the first as a text book of the subject. . . . Mr. Clarke has added immensely to the value of the book."—*Engineer.*

Levelling.

A TREATISE ON THE PRINCIPLES AND PRACTICE OF LEVELLING. Showing its Application to purposes of Railway and Civil Engineering, in the Construction of Roads; with Mr. TELFORD's Rules for the same. By FREDERICK W. SIMMS, F.G.S., M.Inst.C.E. Seventh Edition, with the addition of LAW's Practical Examples for Setting-out Railway Curves, and TRAUTWINE's Field Practice of Laying-out Circular Curves. With 7 Plates and numerous Woodcuts, 8vo, 8s. 6d. cloth. *⁎* TRAUTWINE on Curves may be had separate, 5s.

"The text-book on levelling in most of our engineering schools and colleges."—*Engineer.*
"The publishers have rendered a substantial service to the profession, especially to the younger members, by bringing out the present edition of Mr. Simms's useful work."—*Engineering.*

Heat, Expansion by.

EXPANSION OF STRUCTURES BY HEAT. By JOHN KEILY, C.E., late of the Indian Public Works and Victorian Railway Departments. Crown 8vo, 3s. 6d. cloth.

SUMMARY OF CONTENTS.

Section   I. FORMULAS AND DATA.
Section II. METAL BARS.
Section III. SIMPLE FRAMES.
Section IV. COMPLEX FRAMES AND PLATES.
Section V. THERMAL CONDUCTIVITY.
Section VI. MECHANICAL FORCE OF HEAT.
Section VII. WORK OF EXPANSION AND CONTRACTION.
Section VIII. SUSPENSION BRIDGES.
Section IX. MASONRY STRUCTURES.

"The aim the author has set before him, viz., to show the effects of heat upon metallic and other structures, is a laudable one, for this is a branch of physics upon which the engineer or architect can find but little reliable and comprehensive data in books."—*Builder.*
"Whoever is concerned to know the effect of changes of temperature on such structures as suspension bridges and the like, could not do better than consult Mr. Keily's valuable and handy exposition of the geometrical principles involved in these changes."—*Scotsman.*

Practical Mathematics.

MATHEMATICS FOR PRACTICAL MEN: Being a Commonplace Book of Pure and Mixed Mathematics. Designed chiefly for the use of Civil Engineers, Architects and Surveyors. By OLINTHUS GREGORY, LL.D., F.R.A.S., Enlarged by HENRY LAW, C.E. 4th Edition, carefully Revised by J. R. YOUNG, formerly Professor of Mathematics, Belfast College. With 13 Plates, 8vo, £1 1s. cloth.

"The engineer or architect will here find ready to his hand rules for solving nearly every mathematical difficulty that may arise in his practice. The rules are in all cases explained by means of examples, in which every step of the process is clearly worked out."—*Builder.*
"One of the most serviceable books for practical mechanics. . . It is an instructive book for the student, and a text-book for him who, having once mastered the subjects it treats of, needs occasionally to refresh his memory upon them."—*Building News.*

Hydraulic Tables.

HYDRAULIC TABLES, CO-EFFICIENTS, and FORMULÆ for finding the Discharge of Water from Orifices, Notches, Weirs, Pipes, and Rivers. With New Formulæ, Tables, and General Information on Rainfall, Catchment-Basins, Drainage, Sewerage, Water Supply for Towns and Mill Power. By JOHN NEVILLE, Civil Engineer, M.R.I.A. Third Ed., carefully Revised, with considerable Additions. Numerous Illusts. Cr. 8vo, 14s. cloth.

"Alike valuable to students and engineers in practice; its study will prevent the annoyance of avoidable failures, and assist them to select the readiest means of successfully carrying out any given work connected with hydraulic engineering."—*Mining Journal.*
"It is, of all English books on the subject, the one nearest to completeness. . . From the good arrangement of the matter, the clear explanations, and abundance of formulæ, the carefully calculated tables, and, above all, the thorough acquaintance with both theory and construction, which is displayed from first to last, the book w be found to be an acquisition."—*Architect.*

Hydraulics.

HYDRAULIC MANUAL. Consisting of Working Tables and Explanatory Text. Intended as a Guide in Hydraulic Calculations and Field Operations. By LOWIS D'A. JACKSON, Author of "Aid to Survey Practice," "Modern Metrology," &c. Fourth Edition, Enlarged. Large cr. 8vo, 16s. cl.

"The author has had a wide experience in hydraulic engineering and has been a careful observer of the facts which have come under his notice, and from the great mass of material at his command he has constructed a manual which may be accepted as a trustworthy guide to this branch of the engineer's profession. We can heartily recommend this volume to all who desire to be acquainted with the latest development of this important subject."—*Engineering.*
"The standard-work in this department of mechnnics.'—*Scotsman.*
"The most useful feature of this work is its freedom from what is superannuated, and its thorough adoption of recent experiments; the text is, in fact, in great part a short account of the great modern experiments."—*Nature.*

Drainage.

ON THE DRAINAGE OF LANDS, TOWNS AND BUILDINGS. By G. D. DEMPSEY, C.E., Author of "The Practical Railway Engineer," &c. Revised, with large Additions on RECENT PRACTICE IN DRAINAGE ENGINEERING, by D. KINNEAR CLARK, M.Inst.C.E. Author of "Tramways," "A Manual of Rules, Tables, and Data for Engineers," &c. Second Edition. 12mo, 5s. cloth.

"The new matter added to Mr. Dempsey's excellent work is characterised by the comprehensive grasp and accuracy of detail for which the name of Mr. D. K. Clark is a sufficient voucher."—*Athenæum*.

"As a work on recent practice in drainage engineering, the book is to be commended to all who are making that branch of engineering science their special study."—*Iron*.

"A comprehensive manual on drainage engineering, and a useful introduction to the student." *Building News.*

Tramways and their Working.

TRAMWAYS: THEIR CONSTRUCTION AND WORKING. Embracing a Comprehensive History of the System; with an exhaustive Analysis of the various Modes of Traction, including Horse-Power, Steam, Heated Water, and Compressed Air; a Description of the Varieties of Rolling Stock; and ample Details of Cost and Working Expenses: the Progress recently made in Tramway Construction, &c. &c. By D. KINNEAR CLARK, M.Inst.C.E. With over 200 Wood Engravings, and 13 Folding Plates. Two Vols., large crown 8vo, 30s. cloth.

"All interested in tramways must refer to it, as all railway engineers have turned to the author's work 'Railway Machinery.'"—*Engineer*.

"An exhaustive and practical work on tramways, in which the history of this kind of locomotion, and a description and cost of the various modes of laying tramways, are to be found."—*Building News.*

"The best form of rails, the best mode of construction, and the best mechanical appliances are so fairly indicated in the work under review, that any engineer about to construct a tramway will be enabled at once to obtain the practical information which will be of most service to him.'—*Athenæum*.

Oblique Arches.

A PRACTICAL TREATISE ON THE CONSTRUCTION OF OBLIQUE ARCHES. By JOHN HART. Third Edition, with Plates. Imperial 8vo, 8s. cloth.

Curves, Tables for Setting-out.

TABLES OF TANGENTIAL ANGLES AND MULTIPLES *for Setting-out Curves from 5 to 200 Radius.* By ALEXANDER BEAZELEY, M.Inst.C.E. Third Edition. Printed on 48 Cards, and sold in a cloth box, waistcoat-pocket size, 3s. 6d.

"Each table is printed on a small card, which, being placed on the theodolite, leaves the hands free to manipulate the instrument—no small advantage as regards the rapidity of work."—*Engineer*.

"Very handy; a man may know that all his day's work must fa on two of these cards, which he puts into his own card-case, and leaves the rest behind."—*Athenæum*.

Earthwork.

EARTHWORK TABLES. Showing the Contents in Cubic Yards of Embankments, Cuttings, &c., of Heights or Depths up to an average of 80 feet. By JOSEPH BROADBENT, C.E., and FRANCIS CAMPIN, C.E. Crown 8vo, 5s. cloth.

"The way in which accuracy is attained, by a simple division of each cross section into three elements, two in which are constant and one variable, is ingenious."—*Athenæum*.

Tunnel Shafts.

THE CONSTRUCTION OF LARGE TUNNEL SHAFTS: A *Practical and Theoretical Essay.* By J. H. WATSON BUCK, M.Inst.C.E., Resident Engineer, London and North-Western Railway. Illustrated with Folding Plates, royal 8vo, 12s. cloth.

"Many of the methods given are of extreme practical value to the mason; and the observations on the form of arch, the rules for ordering the stone, and the construction of the templates will be found of considerable use. We commend the book to the engineering profession."—*Building News*.

"Will be regarded by civil engineers as of the utmost value, and calculated to save much time and obviate many mistakes."—*Colliery Guardian*.

Girders, Strength of.

GRAPHIC TABLE FOR FACILITATING THE COMPUTATION OF THE WEIGHTS OF WROUGHT IRON AND STEEL GIRDERS, etc., for Parliamentary and other Estimates. By J. H. WATSON BUCK, M.Inst.C.E. On a Sheet, 2s. 6d.

River Engineering.

RIVER BARS: The Causes of their Formation, and their Treatment by "Induced Tidal Scour;" with a Description of the Successful Reduction by this Method of the Bar at Dublin. By I. J. MANN, Assist. Eng. to the Dublin Port and Docks Board. Royal 8vo, 7s. 6d. cloth.

"We recommend all interested in harbour works—and, indeed, those concerned in the improvements of rivers generally—to read Mr. Mann's interesting work on the treatment of river bars."—*Engineer.*

Trusses.

TRUSSES OF WOOD AND IRON. Practical Applications of Science in Determining the Stresses, Breaking Weights, Safe Loads, Scantlings, and Details of Construction, with Complete Working Drawings. By WILLIAM GRIFFITHS, Surveyor, Assistant Master, Tranmere School of Science and Art. Oblong 8vo, 4s. 6d. cloth.

"This handy little book enters so minutely into every detail connected with the construction of roof trusses, that no student need be ignorant of these matters."—*Practical Engineer.*

Railway Working.

SAFE RAILWAY WORKING. A Treatise on Railway Accidents: Their Cause and Prevention; with a Description of Modern Appliances and Systems. By CLEMENT E. STRETTON, C.E., Vice-President and Consulting Engineer, Amalgamated Society of Railway Servants. With Illustrations and Coloured Plates. Second Edition, Enlarged. Crown 8vo, 3s. 6d. cloth. [*Just published.*

"A book for the engineer, the directors, the managers; and, in short, all who wish for information on railway matters will find a perfect encyclopædia in 'Safe Railway Working.'"—*Railway Review.*

"We commend the remarks on railway signalling to all railway managers, especially where a uniform code and practice is advocated."—*Herepath's Railway Journal.*

"The author may be congratulated on having collected, in a very convenient form, much valuable information on the principal questions affecting the safe working of railways."—*Railway Engineer.*

Field-Book for Engineers.

THE ENGINEER'S, MINING SURVEYOR'S, AND CONTRACTOR'S FIELD-BOOK. Consisting of a Series of Tables, with Rules, Explanations of Systems, and use of Theodolite for Traverse Surveying and Plotting the Work with minute accuracy by means of Straight Edge and Set Square only; Levelling with the Theodolite, Casting-out and Reducing Levels to Datum, and Plotting Sections in the ordinary manner; setting-out Curves with the Theodolite by Tangential Angles and Multiples, with Right and Left-hand Readings of the Instrument: Setting-out Curves without Theodolite, on the System of Tangential Angles by sets of Tangents and Offsets; and Earthwork Tables to 80 feet deep, calculated for every 6 inches in depth. By W. DAVIS HASKOLL, C.E. With numerous Woodcuts. Fourth Edition, Enlarged. Crown 8vo, 12s. cloth.

"The book is very handy; the separate tables of sines and tangents to every minute will make it useful for many other purposes, the genuine traverse tables existing all the same."—*Athenæum.*

"Every person engaged in engineering field operations will estimate the importance of such a work and the amount of valuable time which will be saved by reference to a set of reliable tables prepared with the accuracy and fulness of those given in this volume."—*Railway News.*

Earthwork, Measurement of.

A MANUAL ON EARTHWORK. By ALEX. J. S. GRAHAM, C.E. With numerous Diagrams. Second Edition. 18mo, 2s. 6d. cloth.

"A great amount of practical information, very admirably arranged, and available for rough estimates, as well as for the more exact calculations required in the engineer's and contractor's offices."—*Artisan.*

Strains in Ironwork.

THE STRAINS ON STRUCTURES OF IRONWORK; with Practical Remarks on Iron Construction. By F. W. SHEILDS, M.Inst.C.E. Second Edition, with 5 Plates. Royal 8vo, 5s. cloth.

The student cannot find a better little book on this subject."—*Engineer.*

Cast Iron and other Metals, Strength of.

A PRACTICAL ESSAY ON THE STRENGTH OF CAST IRON AND OTHER METALS. By THOMAS TREDGOLD, C.E. Fifth Edition, including HODGKINSON'S Experimental Researches. 8vo, 12s. cloth.

ARCHITECTURE, BUILDING, etc.

Construction.
THE SCIENCE OF BUILDING : An Elementary Treatise on the *Principles of Construction.* By E. WYNDHAM TARN, M.A., Architect. Third Edition, Enlarged, with 59 Engravings. Fcap. 8vo, 4s. cloth.
"A very valuable book, which we strongly recommend to all students."—*Builder.*
"No architectural student should be without this handbook."—*Architect.*

Villa Architecture.
A HANDY BOOK OF VILLA ARCHITECTURE: Being a *Series of Designs for Villa Residences in various Styles.* With Outline Specifications and Estimates. By C. WICKES, Author of "The Spires and Towers of England," &c. 61 Plates, 4to, £1 11s. 6d. half-morocco, gilt edges.
"The whole of the designs bear evidence of their being the work of an artistic architect, and they will prove very valuable and suggestive."—*Building News.*

Text-Book for Architects.
THE ARCHITECT'S GUIDE: Being a Text-Book of Useful *Information for Architects, Engineers, Surveyors, Contractors, Clerks of Works, &c. &c.* By FREDERICK ROGERS, Architect, Author of "Specifications for Practical Architecture," &c. Second Edition, Revised and Enlarged. With numerous Illustrations. Crown 8vo, 6s. cloth.
"As a text-book of useful information for architects, engineers, surveyors, &c., it would be hard to find a handier or more complete little volume."—*Standard.*
"A young architect could hardly have a better guide-book."—*Timber Trades Journal.*

Taylor and Cresy's Rome.
THE ARCHITECTURAL ANTIQUITIES OF ROME. By the late G. L. TAYLOR, Esq., F.R.I.B.A., and EDWARD CRESY, Esq. New Edition, thoroughly Revised by the Rev. ALEXANDER TAYLOR, M.A. (son of the late G. L. Taylor, Esq.), Fellow of Queen's College, Oxford, and Chaplain of Gray's Inn. Large folio, with 130 Plates, half-bound, £3 3s.
"Taylor and Cresy's work has from its first publication been ranked among those professional books which cannot be bettered. . . . It would be difficult to find examples of drawings, even among those of the most painstaking students of Gothic, more thoroughly worked out than are the one hundred and thirty plates in this volume."—*Architect.*

Linear Perspective.
ARCHITECTURAL PERSPECTIVE: The whole Course and Operations of the Draughtsman in Drawing a Large House in Linear Perspective. Illustrated by 39 Folding Plates. By F. O. FERGUSON. Demy 8vo, 3s. 6d. boards. *[Just published.*

Architectural Drawing.
PRACTICAL RULES ON DRAWING, *for the Operative Builder and Young Student in Architecture.* By GEORGE PYNE. With 14 Plates, 4to, 7s. 6d. boards.

Sir Wm. Chambers on Civil Architecture.
THE DECORATIVE PART OF CIVIL ARCHITECTURE. By Sir WILLIAM CHAMBERS, F.R.S. With Portrait, Illustrations, Notes, and an Examination of Grecian Architecture, by JOSEPH GWILT, F.S.A. Revised and Edited by W. H. LEEDS, with a Memoir of the Author. 66 Plates, 4to, 21s. cloth.

House Building and Repairing.
THE HOUSE-OWNER'S ESTIMATOR; or, What will it Cost to Build, Alter, or Repair? A Price Book adapted to the Use of Unprofessional People, as well as for the Architectural Surveyor and Builder. By JAMES D. SIMON, A.R.I.B.A. Edited and Revised by FRANCIS T. W. MILLER, A.R.I.B.A. With numerous Illustrations. Fourth Edition, Revised. Crown 8vo, 3s. 6d. cloth.
"In two years it will repay its cost a hundred times over."—*Field.*

Cottages and Villas.
COUNTRY AND SUBURBAN COTTAGES AND VILLAS How to Plan and Build Them. Containing 33 Plates, with Introduction, General Explanations, and Description of each Plate. By JAMES W. BOGUE, Architect, Author of "Domestic Architecture," &c. 4to, 10s. 6d. cloth.

The New Builder's Price Book, 1892.

LOCKWOOD'S BUILDER'S PRICE BOOK FOR 1892. A Comprehensive Handbook of the Latest Prices and Data for Builders, Architects, Engineers and Contractors. *Re-constructed, Re-written and Further Enlarged.* By FRANCIS T. W. MILLER. 700 closely-printed pages, crown 8vo, 4s. cloth. [*Just published.*

"This book is a very useful one, and should find a place in every English office connected with the building and engineering professions."—*Industries.*

"This Price Book has been set up in new type. . . . Advantage has been taken of the transformation to add much additional information, and the volume is now an excellent book of reference."—*Architect.*

"In its new and revised form this Price Book is what a work of this kind should be—comprehensive, reliable, well arranged, legible and well bound."—*British Architect.*

"A work of established reputation."—*Athenæum.*

"This very useful handbook is well written, exceedingly clear in its explanations and great care has evidently been taken to ensure accuracy."—*Morning Advertiser.*

Designing, Measuring, and Valuing.

THE STUDENT'S GUIDE to the PRACTICE of MEASURING AND VALUING ARTIFICERS' WORKS. Containing Directions for taking Dimensions, Abstracting the same, and bringing the Quantities into Bill, with Tables of Constants for Valuation of Labour, and for the Calculation of Areas and Solidities. Originally edited by EDWARD DOBSON, Architect. With Additions on Mensuration and Construction, and a New Chapter on Dilapidations, Repairs, and Contracts, by E. WYNDHAM TARN, M.A. Sixth Edition, including a Complete Form of a Bill of Quantities. With 8 Plates and 63 Woodcuts. Crown 8vo, 7s. 6d. cloth.

"Well fulfils the promise of its title-page, and we can thoroughly recommend it to the class for whose use it has been compiled. Mr. Tarn's additions and revisions have much increased the usefulness of the work, and have especially augmented its value to students."—*Engineering.*

"This edition will be found the most complete treatise on the principles of measuring and valuing artificers' work that has yet been published."—*Building News.*

Pocket Estimator and Technical Guide.

THE POCKET TECHNICAL GUIDE, MEASURER AND ESTIMATOR FOR BUILDERS AND SURVEYORS. Containing Technical Directions for Measuring Work in all the Building Trades, Complete Specifications for Houses, Roads, and Drains, and an easy Method of Estimating the parts of a Building collectively. By A. C. BEATON, Author of "Quantities and Measurements," &c. Sixth Edition, Revised. With 53 Woodcuts, waistcoat-pocket size, 1s. 6d. gilt edges. [*Just published.*

"No builder, architect, surveyor, or valuer should be without his 'Beaton.'"—*Building News.*

"Contains an extraordinary amount of information in daily requisition in measuring and estimating. Its presence in the pocket will save valuable time and trouble."—*Building World.*

Donaldson on Specifications.

THE HANDBOOK OF SPECIFICATIONS; or, Practical Guide to the Architect, Engineer, Surveyor, and Builder, in drawing up Specifications and Contracts for Works and Constructions. Illustrated by Precedents of Buildings actually executed by eminent Architects and Engineers. By Professor T. L. DONALDSON, P.R.I.B.A., &c. New Edition, in One large Vol., 8vo, with upwards of 1,000 pages of Text, and 33 Plates, £1 11s. 6d. cloth.

"In this work forty-four specifications of executed works are given, including the specifications for parts of the new Houses of Parliament, by Sir Charles Barry, and for the new Royal Exchange, by Mr. Tite, M.P. The latter, in particular, is a very complete and remarkable document. It embodies, to a great extent, as Mr. Donaldson mentions, 'the bill of quantities with the description of the works.' . . . It is valuable as a record, and more valuable still as a book of precedents. . . . Suffice it to say that Donaldson's 'Handbook of Specifications must be bought by all architects."—*Builder.*

Bartholomew and Rogers' Specifications.

SPECIFICATIONS FOR PRACTICAL ARCHITECTURE. A Guide to the Architect, Engineer, Surveyor, and Builder. With an Essay on the Structure and Science of Modern Buildings. Upon the Basis of the Work by ALFRED BARTHOLOMEW, thoroughly Revised, Corrected, and greatly added to by FREDERICK ROGERS, Architect. Second Edition, Revised, with Additions. With numerous Illustrations, medium 8vo, 15s. cloth.

"The collection of specifications prepared by Mr. Rogers on the basis of Bartholomew's work is too well known to need any recommendation from us. It is one of the books with which every young architect must be equipped; for time has shown that the specifications cannot be set aside through any defect in them."—*Architect.*

Building; Civil and Ecclesiastical.

A BOOK ON BUILDING, Civil *and* Ecclesiastical, including Church Restoration ; with the Theory of Domes and the Great Pyramid, &c. By Sir EDMUND BECKETT, Bart., LL.D., F.R.A.S., Author of "Clocks and Watches, and Bells," &c. Second Edition, Enlarged. Fcap. 8vo, 5s. cloth.
"A book which is always amusing and nearly always instructive. The style throughout is in the highest degree condensed and epigrammatic."—*Times.*

Ventilation of Buildings.

VENTILATION. A *Text Book to the Practice of the Art of Ventilating Buildings.* With a Chapter upon Air Testing. By W. P. BUCHAN, R.P., Sanitary and Ventilating Engineer, Author of "Plumbing," &c. With 170 Illustrations. 12mo, 4s. cloth boards. [*Just published.*

The Art of Plumbing.

PLUMBING. *A Text Book to the Practice of the Art or Craft of the Plumber, with Supplementary Chapters on House Drainage, embodying the latest Improvements.* By WILLIAM PATON BUCHAN, R.P., Sanitary Engineer and Practical Plumber. Sixth Edition, Enlarged to 370 pages, and 380 Illustrations. 12mo, 4s. cloth boards.
"A text book which may be safely put in the hands of every young plumber, and which will also be found useful by architects and medical professors."—*Builder.*
"A valuable text book, and the only treatise which can be regarded as a really reliable manual of the plumber's art."—*Building News.*

Geometry for the Architect, Engineer, etc.

PRACTICAL GEOMETRY, *for the Architect, Engineer and Mechanic.* Giving Rules for the Delineation and Application of various Geometrical Lines, Figures and Curves. By E. W. TARN, M.A., Architect, Author of "The Science of Building," &c. Second Edition. With 172 Illustrations, demy 8vo, 9s. cloth.
"No book with the same objects in view has ever been published in which the clearness of the rules laid down and the illustrative diagrams have been so satisfactory."—*Scotsman.*

The Science of Geometry.

THE GEOMETRY OF COMPASSES; *or, Problems Resolved by the mere Description of Circles, and the use of Coloured Diagrams and Symbols.* By OLIVER BYRNE. Coloured Plates. Crown 8vo, 3s. 6d. cloth.
"The treatise is a good one, and remarkable—like all Mr. Byrne's contributions to the science of geometry—for the lucid character of its teaching."—*Building News.*

DECORATIVE ARTS, etc.

Woods and Marbles (Imitation of).

SCHOOL OF PAINTING FOR THE IMITATION OF WOODS AND MARBLES, as Taught and Practised by A. R. VAN DER BURG and P. VAN DER BURG, Directors of the Rotterdam Painting Institution. Royal folio, 18¾ by 12⅜ in., Illustrated with 24 full-size Coloured Plates; also 12 plain Plates, comprising 154 Figures. Second and Cheaper Edition. Price £1 11s. 6d.

List of Plates.

1. Various Tools required for Wood Painting—2, 3. Walnut: Preliminary Stages of Graining and Finished Specimen—4. Tools used for Marble Painting and Method of Manipulation—5, 6. St. Remi Marble: Earlier Operations and Finished Specimen—7. Methods of Sketching different Grains, Knots, &c.—8, 9. Ash: Preliminary Stages and Finished Specimen—10. Methods of Sketching Marble Grains—11, 12. Breche Marble: Preliminary Stages of Working and Finished Specimen—13. Maple: Methods of Producing the different Grains—14, 15. Bird's-eye Maple: Preliminary Stages and Finished Specimen—16. Methods of Sketching the different Species of White Marble—17, 18. White Marble: Preliminary Stages of Process and Finished Specimen—19. Mahogany: Specimens of various Grains and Methods of Manipulation—20, 21. Mahogany: Earlier Stages and Finished Specimen—22, 23, 24. Sienna Marble: Varieties of Grain, Preliminary Stages and Finished Specimen—25, 26, 27. Juniper Wood: Methods of producing Grain, &c.: Preliminary Stages and Finished Specimen—28, 29, 30. Vert de Mer Marble: Varieties of Grain and Methods of Working Unfinished and Finished Specimens—31, 32, 33. Oak: Varieties of Grain, Tools Employed, and Methods of Manipulation, Preliminary Stages and Finished Specimen—34, 35, 36. Waulsort Marble: Varieties of Grain, Unfinished and Finished Specimens.

⁎⁎ OPINIONS OF THE PRESS.
"Those who desire to attain skill in the art of painting woods and marbles will find advantage in consulting this book. . . . Some of the Working Men's Clubs should give their young men the opportunity to study it."—*Builder.*
"A comprehensive guide to the art. The explanations of the processes, the manipulation and management of the colours, and the beautifully executed plates will not be the least valuable to the student who aims at making his work a faithful transcript of nature."—*Building News.*

House Decoration.

ELEMENTARY DECORATION. A Guide to the Simpler Forms of Everyday Art, as applied to the Interior and Exterior Decoration of Dwelling Houses, &c. By JAMES W. FACEY, Jun. With 68 Cuts. 12mo, 2s. cloth limp.

PRACTICAL HOUSE DECORATION: A Guide to the Art of Ornamental Painting, the Arrangement of Colours in Apartments, and the principles of Decorative Design. With some Remarks upon the Nature and Properties of Pigments. By JAMES WILLIAM FACEY, Author of "Elementary Decoration," &c. With numerous Illustrations. 12mo, 2s. 6d. cloth limp.

N.B.—The above Two Works together in One Vol., strongly half-bound, 5s.

Colour.

A GRAMMAR OF COLOURING. Applied to Decorative Painting and the Arts. By GEORGE FIELD. New Edition, Revised, Enlarged, and adapted to the use of the Ornamental Painter and Designer. By ELLIS A. DAVIDSON. With New Coloured Diagrams and Engravings. 12mo, 3s. 6d. cloth boards.

"The book is a most useful *resume* of the properties of pigments."—*Builder.*

House Painting, Graining, etc.

HOUSE PAINTING, GRAINING, MARBLING, AND SIGN WRITING, A Practical Manual of. By ELLIS A. DAVIDSON. Sixth Edition. With Coloured Plates and Wood Engravings. 12mo, 6s. cloth boards.

"A mass of information, of use to the amateur and of value to the practical man."—*English Mechanic.*

"Simply invaluable to the youngster entering upon this particular calling, and highly serviceable to the man who is practising it."—*Furniture Gazette.*

Decorators, Receipts for.

THE DECORATOR'S ASSISTANT: A Modern Guide to Decorative Artists and Amateurs, Painters, Writers, Gilders, &c. Containing upwards of 600 Receipts, Rules and Instructions; with a variety of Information for General Work connected with every Class of Interior and Exterior Decorations, &c. Fourth Edition, Revised. 152 pp., crown 8vo, 1s. in wrapper.

"Full of receipts of value to decorators, painters, gilders, &c. The book contains the gist of larger treatises on colour and technical processes. It would be difficult to meet with a work so full of varied information on the painter's art."—*Building News.*

"We recommend the work to all who, whether for pleasure or profit, require a guide to decoration."—*Plumber and Decorator.*

Moyr Smith on Interior Decoration.

ORNAMENTAL INTERIORS, ANCIENT AND MODERN. By J. MOYR SMITH. Super-royal 8vo, with 32 full-page Plates and numerous smaller Illustrations, handsomely bound in cloth, gilt top, price 18s.

"The book is well illustrated and handsomely got up, and contains some true criticism and a good many good examples of decorative treatment."—*The Builder.*

"This is the most elaborate and beautiful work on the artistic decoration of interiors that we have seen. . . . The scrolls, panels and other designs from the author's own pen are very beautiful and chaste; but he takes care that the designs of other men shall figure even more than his own."—*Liverpool Albion.*

"To all who take an interest in elaborate domestic ornament this handsome volume will be welcome."—*Graphic.*

British and Foreign Marbles.

MARBLE DECORATION and the Terminology of British and Foreign Marbles. A Handbook for Students. By GEORGE H. BLAGROVE, Author of "Shoring and its Application," &c. With 28 Illustrations. Crown 8vo, 3s. 6d. cloth.

"This most useful and much wanted handbook should be in the hands of every architect and builder."—*Building World.*

"It is an excellent manual for students, and interesting to artistic readers generally."—*Saturday Review.*

"A carefully and usefully written treatise; the work is essentially practical."—*Scotsman.*

Marble Working, etc.

MARBLE AND MARBLE WORKERS: A Handbook for Architects, Artists, Masons and Students. By ARTHUR LEE, Author of "A Visit to Carrara," "The Working of Marble," &c. Small crown 8vo, 2s. cloth.

"A really valuable addition to the technical literature of architects and masons."—*Building News.*

DELAMOTTE'S WORKS ON ILLUMINATION AND ALPHABETS.

A PRIMER OF THE ART OF ILLUMINATION, for the Use of Beginners: with a Rudimentary Treatise on the Art, Practical Directions for its exercise, and Examples taken from Illuminated MSS., printed in Gold and Colours. By F. DELAMOTTE. New and Cheaper Edition. Small 4to, 6s. ornamental boards.

"The examples of ancient MSS. recommended to the student, which, with much good sense, the author chooses from collections accessible to all, are selected with judgment and knowledge, as well as taste."—*Athenæum.*

ORNAMENTAL ALPHABETS, Ancient and Mediæval, from the Eighth Century, with Numerals; including Gothic, Church-Text, large and small, German, Italian, Arabesque, Initials for Illumination, Monograms, Crosses, &c. &c., for the use of Architectural and Engineering Draughtsmen, Missal Painters, Masons, Decorative Painters, Lithographers, Engravers, Carvers, &c. &c. Collected and Engraved by F. DELAMOTTE, and printed in Colours. New and Cheaper Edition. Royal 8vo, oblong, 2s. 6d. ornamental boards.

"For those who insert enamelled sentences round gilded chalices, who blazon shop legends over shop-doors, who letter church walls with pithy sentences from the Decalogue, this book will be useful."—*Athenæum.*

EXAMPLES OF MODERN ALPHABETS, Plain and Ornamental; including German, Old English, Saxon, Italic, Perspective, Greek, Hebrew, Court Hand, Engrossing, Tuscan, Riband, Gothic, Rustic, and Arabesque; with several Original Designs, and an Analysis of the Roman and Old English Alphabets, large and small, and Numerals, for the use of Draughtsmen, Surveyors, Masons, Decorative Painters, Lithographers, Engravers, Carvers, &c. Collected and Engraved by F. DELAMOTTE, and printed in Colours. New and Cheaper Edition. Royal 8vo, oblong, 2s. 6d. ornamental boards.

"There is comprised in it every possible shape into which the letters of the alphabet and numerals can be formed, and the talent which has been expended in the conception of the various plain and ornamental letters is wonderful."—*Standard.*

MEDIÆVAL ALPHABETS AND INITIALS FOR ILLUMINATORS. By F. G. DELAMOTTE. Containing 21 Plates and Illuminated Title, printed in Gold and Colours. With an Introduction by J. WILLIS BROOKS. Fourth and Cheaper Edition. Small 4to, 4s. ornamental boards.

"A volume in which the letters of the alphabet come forth glorified in gilding and all the colours of the prism interwoven and intertwined and intermingled."—*Sun.*

THE EMBROIDERER'S BOOK OF DESIGN. Containing Initials, Emblems, Cyphers, Monograms, Ornamental Borders, Ecclesiastical Devices, Mediæval and Modern Alphabets, and National Emblems. Collected by F. DELAMOTTE, and printed in Colours. Oblong royal 8vo, 1s. 6d. ornamental wrapper.

"The book will be of great assistance to ladies and young children who are endowed with th art of plying the needle in this most ornamental and useful pretty work."—*East Anglian Times.*

Wood Carving.

INSTRUCTIONS IN WOOD-CARVING, for Amateurs; with Hints on Design. By A LADY. With Ten Plates. New and Cheaper Edition. Crown 8vo, 2s. in emblematic wrapper.

"The handicraft of the wood-carver, so well as a book can impart it, may be learnt from 'A Lady's' publication."—*Athenæum.*
"The directions given are plain and easily understood."—*English Mechanic.*

Glass Painting.

GLASS STAINING AND THE ART OF PAINTING ON GLASS. From the German of Dr. GESSERT and EMANUEL OTTO FROMBERG. With an Appendix on THE ART OF ENAMELLING. 12mo, 2s. 6d. cloth limp.

Letter Painting.

THE ART OF LETTER PAINTING MADE EASY. By JAMES GREIG BADENOCH. With 12 full-page Engravings of Examples, 1s. 6d. cloth limp.

"The system is a simple one, but quite original, and well worth the careful attention of letter painters. It can be easily mastered and remembered."—*Building News.*

CARPENTRY, TIMBER, etc.

Tredgold's Carpentry, Revised & Enlarged by Tarn.
THE ELEMENTARY PRINCIPLES OF CARPENTRY. A Treatise on the Pressure and Equilibrium of Timber Framing, the Resistance of Timber, and the Construction of Floors, Arches, Bridges, Roofs, Uniting Iron and Stone with Timber, &c. To which is added an Essay on the Nature and Properties of Timber, &c., with Descriptions of the kinds of Wood used in Building; also numerous Tables of the Scantlings of Timber for different purposes, the Specific Gravities of Materials, &c. By THOMAS TREDGOLD, C.E. With an Appendix of Specimens of Various Roofs of Iron and Stone, Illustrated. Seventh Edition, thoroughly revised and considerably enlarged by E. WYNDHAM TARN, M.A., Author of "The Science of Building," &c. With 61 Plates, Portrait of the Author, and several Woodcuts. In one large vol., 4to, price £1 5s. cloth.

"Ought to be in every architect's and every builder's library."—*Builder.*
"A work whose monumental excellence must commend it wherever skilful carpentry is concerned. The author's principles are rather confirmed than impaired by time. The additional plates are of great intrinsic value."—*Building News.*

Woodworking Machinery.
WOODWORKING MACHINERY: Its Rise, Progress, and Construction. With Hints on the Management of Saw Mills and the Economical Conversion of Timber. Illustrated with Examples of Recent Designs by leading English, French, and American Engineers. By M. POWIS BALE, A.M.Inst.C.E., M.I.M.E. Large crown 8vo, 12s. 6d. cloth.

"Mr. Bale is evidently an expert on the subject and he has collected so much information that his book is all-sufficient for builders and others engaged in the conversion of timber."—*Architect.*
"The most comprehensive compendium of wood-working machinery we have seen. The author is a thorough master of his subject."—*Building News.*
"The appearance of this book at the present time will, we should think, give a considerable impetus to the onward march of the machinist engaged in the designing and manufacture of wood-working machines. It should be in the office of every wood-working factory."—*English Mechanic.*

Saw Mills.
SAW MILLS: Their Arrangement and Management, and the Economical Conversion of Timber. (A Companion Volume to "Woodworking Machinery.") By M. POWIS BALE. With numerous Illustrations. Crown 8vo, 10s. 6d. cloth.

"The *administration* of a large sawing establishment is discussed, and the subject examined from a financial standpoint. We could not desire a more complete or practical treatise."—*Builder.*
"We highly recommend Mr. Bale's work to the attention and perusal of all those who are engaged in the art of wood conversion, or who are about building or remodelling saw-mills on improved principles."—*Building News.*

Carpentering.
THE CARPENTER'S NEW GUIDE; or, Book of Lines for Carpenters; comprising all the Elementary Principles essential for acquiring a knowledge of Carpentry. Founded on the late PETER NICHOLSON'S Standard Work. A New Edition, Revised by ARTHUR ASHPITEL, F.S.A. Together with Practical Rules on Drawing, by GEORGE PYNE. With 74 Plates, 4to, £1 1s. cloth.

Handrailing and Stairbuilding.
A PRACTICAL TREATISE ON HANDRAILING: Showing New and Simple Methods for Finding the Pitch of the Plank, Drawing the Moulds, Bevelling, Jointing-up, and Squaring the Wreath. By GEORGE COLLINGS. Second Edition, Revised and Enlarged, to which is added A TREATISE ON STAIRBUILDING. With Plates and Diagrams. 12mo, 2s. 6d. cloth limp.

"Will be found of practical utility in the execution of this difficult branch of joinery."—*Builder.*
"Almost every difficult phase of this somewhat intricate branch of joinery is elucidated by the aid of plates and explanatory letterpress."—*Furniture Gazette.*

Circular Work.
CIRCULAR WORK IN CARPENTRY AND JOINERY: A Practical Treatise on Circular Work of Single and Double Curvature. By GEORGE COLLINGS, Author of "A Practical Treatise on Handrailing." Illustrated with numerous Diagrams. Second Edition. 12mo, 2s. 6d. cloth limp.

"An excellent example of what a book of this kind should be. Cheap in price, clear in definition and practical in the examples selected."—*Builder.*

Timber Merchant's Companion.

THE TIMBER MERCHANT'S AND BUILDER'S COMPANION. Containing New and Copious Tables of the Reduced Weight and Measurement of Deals and Battens, of all sizes, from One to a Thousand Pieces, and the relative Price that each size bears per Lineal Foot to any given Price per Petersburg Standard Hundred; the Price per Cube Foot of Square Timber to any given Price per Load of 50 Feet; the proportionate Value of Deals and Battens by the Standard, to Square Timber by the Load of 50 Feet; the readiest mode of ascertaining the Price of Scantling per Lineal Foot of any size, to any given Figure per Cube Foot, &c. &c. By WILLIAM DOWSING. Fourth Edition, Revised and Corrected. Cr. 8vo, 3s. cl.

"We are glad to see a fourth edition of these admirable tables, which for correctness and simplicity of arrangement leave nothing to be desired."—*Timber Trades Journal.*

"An exceedingly well-arranged, clear, and concise manual of tables for the use of all who buy or sell timber."—*Journal of Forestry.*

Practical Timber Merchant.

THE PRACTICAL TIMBER MERCHANT. Being a Guide for the use of Building Contractors, Surveyors, Builders, &c., comprising useful Tables for all purposes connected with the Timber Trade, Marks of Wood, Essay on the Strength of Timber, Remarks on the Growth of Timber, &c. By W. RICHARDSON. Fcap. 8vo, 3s. 6d. cloth.

"This handy manual contains much valuable information for the use of timber merchants, builders, foresters, and all others connected with the growth, sale, and manufacture of timber."—*Journal of Forestry.*

Timber Freight Book.

THE TIMBER MERCHANT'S, SAW MILLER'S, AND IMPORTER'S FREIGHT BOOK AND ASSISTANT. Comprising Rules, Tables, and Memoranda relating to the Timber Trade. By WILLIAM RICHARDSON, Timber Broker; together with a Chapter on "SPEEDS OF SAW MILL MACHINERY," by M. POWIS BALE, M.I.M.E., &c. 12mo, 3s. 6d. cl. boards.

"A very useful manual of rules, tables, and memoranda relating to the timber trade. We recommend it as a compendium of calculation to all timber measurers and merchants, and as supplying a real want in the trade."—*Building News.*

Packing-Case Makers, Tables for.

PACKING-CASE TABLES; showing the number of Superficial Feet in Boxes or Packing-Cases, from six inches square and upwards. By W. RICHARDSON, Timber Broker. Third Edition. Oblong 4to, 3s. 6d. cl.

"Invaluable labour-saving tables."—*Ironmonger.*

"Will save much labour and calculation."—*Grocer.*

Superficial Measurement.

THE TRADESMAN'S GUIDE TO SUPERFICIAL MEASUREMENT. Tables calculated from 1 to 200 inches in length, by 1 to 108 inches in breadth. For the use of Architects, Surveyors, Engineers, Timber Merchants, Builders, &c. By JAMES HAWKINGS. Third Edition. Fcap., 3s. 6d. cloth.

"A useful collection of tables to facilitate rapid calculation of surfaces. The exact area of any surface of which the limits have been ascertained can be instantly determined. The book will be found of the greatest utility to all engaged in building operations."—*Scotsman.*

"These tables will be found of great assistance to all who require to make calculations in superficial measurement."—*English Mechanic.*

Forestry.

THE ELEMENTS OF FORESTRY. Designed to afford Information concerning the Planting and Care of Forest Trees for Ornament or Profit, with Suggestions upon the Creation and Care of Woodlands. By F. B. HOUGH. Large crown 8vo, 10s. cloth.

Timber Importer's Guide.

THE TIMBER IMPORTER'S, TIMBER MERCHANT'S AND BUILDER'S STANDARD GUIDE. By RICHARD E. GRANDY. Comprising an Analysis of Deal Standards, Home and Foreign, with Comparative Values and Tabular Arrangements for fixing Nett Landed Cost on Baltic and North American Deals, including all intermediate Expenses, Freight, Insurance, &c. &c. Together with copious Information for the Retailer and Builder. Third Edition, Revised. 12mo, 2s. cloth limp.

"Everything it pretends to be: built up gradually, it leads one from a forest to a treenall, and throws in, as a makeweight, a host of material concerning bricks, columns, cisterns, &c."—*English Mechanic.*

MARINE ENGINEERING, NAVIGATION, etc.

Chain Cables.
CHAIN CABLES AND CHAINS. Comprising Sizes and Curves of Links, Studs, &c., Iron for Cables and Chains, Chain Cable and Chain Making, Forming and Welding Links, Strength of Cables and Chains, Certificates for Cables, Marking Cables, Prices of Chain Cables and Chains, Historical Notes, Acts of Parliament, Statutory Tests, Charges for Testing, List of Manufacturers of Cables, &c. &c. By THOMAS W. TRAILL, F.E.R.N., M. Inst. C.E., Engineer Surveyor in Chief, Board of Trade, Inspector of Chain Cable and Anchor Proving Establishments, and General Superintendent, Lloyd's Committee on Proving Establishments. With numerous Tables, Illustrations and Lithographic Drawings. Folio, £2 2s. cloth, bevelled boards.

"It contains a vast amount of valuable information. Nothing seems to be wanting to make it a complete and standard work of reference on the subject."—*Nautical Magazine.*

Marine Engineering.
MARINE ENGINES AND STEAM VESSELS (A Treatise on). By ROBERT MURRAY, C.E. Eighth Edition, thoroughly Revised, with considerable Additions by the Author and by GEORGE CARLISLE, C.E., Senior Surveyor to the Board of Trade at Liverpool. 12mo, 5s. cloth boards.

"Well adapted to give the young steamship engineer or marine engine and boiler maker a general introduction into his practical work."—*Mechanical World.*

"We feel sure that this thoroughly revised edition will continue to be as popular in the future as it has been in the past, as, for its size, it contains more useful information than any similar treatise."—*Industries.*

"The information given is both sound and sensible, and well qualified to direct young sea-going hands on the straight road to the extra chief's certificate. Most useful to surveyors, inspectors, draughtsmen, and all young engineers who take an interest in their profession."—*Glasgow Herald.*

"An indispensable manual for the student of marine engineering."—*Liverpool Mercury.*

Pocket-Book for Naval Architects and Shipbuilders.
THE NAVAL ARCHITECT'S AND SHIPBUILDER'S POCKET-BOOK of Formulæ, Rules, and Tables, and MARINE ENGINEER'S AND SURVEYOR'S Handy Book of Reference. By CLEMENT MACKROW, Member of the Institution of Naval Architects, Naval Draughtsman. Fourth Edition, Revised. With numerous Diagrams, &c. Fcap., 12s. 6d. strongly bound in leather.

"Will be found to contain the most useful tables and formulæ required by shipbuilders, carefully collected from the best authorities, and put together in a popular and simple form."—*Engineer.*

"The professional shipbuilder has now, in a convenient and accessible form, reliable data for solving many of the numerous problems that present themselves in the course of his work."—*Iron.*

"There is scarcely a subject on which a naval architect or shipbuilder can require to refresh his memory which will not be found within the covers of Mr. Mackrow's book."—*English Mechanic.*

Pocket-Book for Marine Engineers.
A POCKET-BOOK OF USEFUL TABLES AND FORMULÆ FOR MARINE ENGINEERS. By FRANK PROCTOR, A.I.N.A. Third Edition. Royal 32mo, leather, gilt edges, with strap, 4s.

"We recommend it to our readers as going far to supply a long-felt want."—*Naval Science.*

"A most useful companion to all marine engineers."—*United Service Gazette.*

Introduction to Marine Engineering.
ELEMENTARY ENGINEERING: A Manual for Young Marine Engineers and Apprentices. In the Form of Questions and Answers on Metals, Alloys, Strength of Materials, Construction and Management of Marine Engines and Boilers, Geometry, &c. &c. With an Appendix of Useful Tables. By JOHN SHERREN BREWER, Government Marine Surveyor, Hongkong. Small crown 8vo, 2s. cloth.

"Contains much valuable information for the class for whom it is intended, especially in the chapters on the management of boilers and engines."—*Nautical Magazine.*

"A useful introduction to the more elaborate text books."—*Scotsman.*

"To a student who has the requisite desire and resolve to attain a thorough knowledge, Mr. Brewer offers decidedly useful help."—*Athenæum.*

Navigation.
PRACTICAL NAVIGATION. Consisting of THE SAILOR'S SEA-BOOK, by JAMES GREENWOOD and W. H. ROSSER; together with the requisite Mathematical and Nautical Tables for the Working of the Problems, by HENRY LAW, C.E., and Professor J. R. YOUNG. Illustrated. 12mo, 7s. strongly half-bound.

MINING AND METALLURGY.

Metalliferous Mining in the United Kingdom.

BRITISH MINING: A Treatise on the History, Discovery, Practical Development, and Future Prospects of Metalliferous Mines in the United Kingdom. By ROBERT HUNT, F.R.S., Keeper of Mining Records; Editor of "Ure's Dictionary of Arts, Manufactures, and Mines," &c. Upwards of 950 pp., with 230 Illustrations. Second Edition, Revised. Super-royal 8vo, £2 2s. cloth.

"One of the most valuable works of reference of modern times. Mr. Hunt, as keeper of mining records of the United Kingdom, has had opportunities for such a task not enjoyed by anyone else, and has evidently made the most of them. . . . The language and style adopted are good, and the treatment of the various subjects laborious, conscientious, and scientific."—*Engineering.*

"The book is, in fact, a treasure-house of statistical information on mining subjects, and we knew of no other work embodying so great a mass of matter of this kind. Were this the only merit of Mr. Hunt's volume, it would be sufficient to render it indispensable in this library of everyone interested in the development of the mining and metallurgical industries of this country."—*Athenæum.*

"A mass of information not elsewhere available, and of the greatest value to those who may be interested in our great mineral industries."—*Engineer.*

"A sound, business-like collection of interesting facts. . . . The amount of information Mr. Hunt has brought together is enormous. . . . The volume appears likely to convey more instruction upon the subject than any work hitherto published."—*Mining Journal.*

Colliery Management.

THE COLLIERY MANAGER'S HANDBOOK: A Comprehensive Treatise on the Laying-out and Working of Collieries, Designed as a Book of Reference for Colliery Managers, and for the Use of Coal-Mining Students preparing for First-class Certificates. By CALEB PAMELY, Mining Engineer and Surveyor; Member of the North of England Institute of Mining and Mechanical Engineers; and Member of the South Wales Institute of Mining Engineers. With nearly 500 Plans, Diagrams, and other Illustrations. Medium 8vo, about 600 pages. Price £1 5s. strongly bound.
[*Just published.*

Coal and Iron.

THE COAL AND IRON INDUSTRIES OF THE UNITED KINGDOM. Comprising a Description of the Coal Fields, and of the Principal Seams of Coal, with Returns of their Produce and its Distribution, and Analyses of Special Varieties. Also an Account of the occurrence of Iron Ores in Veins or Seams; Analyses of each Variety; and a History of the Rise and Progress of Pig Iron Manufacture. By RICHARD MEADE, Assistant Keeper of Mining Records. With Maps. 8vo, £1 8s. cloth.

"The book is one which must find a place on the shelves of all interested in coal and iron production, and in the iron, steel, and other metallurgical industries."—*Engineer.*

"Of this book we may unreservedly say that it is the best of its class which we have ever met. . . . A book of reference which no one engaged in the iron or coal trades should omit from his library."—*Iron and Coal Trades Review.*

Prospecting for Gold and other Metals.

THE PROSPECTOR'S HANDBOOK: A Guide for the Prospector and Traveller in Search of Metal-Bearing or other Valuable Minerals. By J. W. ANDERSON, M.A. (Camb.), F.R.G.S., Author of "Fiji and New Caledonia." Fifth Edition, thoroughly Revised and Enlarged. Small crown 8vo, 3s. 6d. cloth.

"Will supply a much felt want, especially among Colonists, in whose way are so often thrown many mineralogical specimens the value of which it is difficult to determine."—*Engineer.*

"How to find commercial minerals, and how to identify them when they are found, are the leading points to which attention is directed. The author has managed to pack as much practical detail into his pages as would supply material for a book three times its size."—*Mining Journal.*

Mining Notes and Formulæ.

NOTES AND FORMULÆ FOR MINING STUDENTS. By JOHN HERMAN MERIVALE, M.A., Certificated Colliery Manager, Professor of Mining in the Durham College of Science, Newcastle-upon-Tyne. Third Edition, Revised and Enlarged. Small crown 8vo, 2s. 6d. cloth.

"Invaluable to anyone who is working up for an examination on mining subjects."—*Coal and Iron Trades Review.*

"The author has done his work in an exceedingly creditable manner, and has produced a book that will be of service to students, and those who are practically engaged in mining operations."—*Engineer.*

"A vast amount of technical matter of the utmost value to mining engineers, and of considerable interest to students."—*Schoolmaster.*

Explosives.

A HANDBOOK ON MODERN EXPLOSIVES. Being a Practical Treatise on the Manufacture and Application of Dynamite, Gun-Cotton, Nitro-Glycerine and other Explosive Compounds. Including the Manufacture of Collodion-Cotton. By M. EISSLER, Mining Engineer and Metallurgical Chemist, Author of "The Metallurgy of Gold," &c. With about 100 Illustrations. Crown 8vo, 10s. 6d. cloth.

"Useful not only to the miner, but also to officers of both services to whom blasting and the use of explosives generally may at any time become a necessary auxiliary."—*Nature.*
"A veritable mine of information on the subject of explosives employed for military, mining and blasting purposes."—*Army and Navy Gazette.*
"The book is clearly written. Taken as a whole, we consider it an excellent little book and one that should be found of great service to miners and others who are engaged in work requiring the use of explosives."—*Athenæum.*

Gold, Metallurgy of.

THE METALLURGY OF GOLD: A Practical Treatise on the Metallurgical Treatment of Gold-bearing Ores. Including the Processes of Concentration and Chlorination, and the Assaying, Melting and Refining of Gold. By M. EISSLER, Mining Engineer and Metallurgical Chemist, formerly Assistant Assayer of the U.S. Mint, San Francisco. Third Edition, Revised and greatly Enlarged. With 187 Illustrations. Crown 8vo, 12s. 6d. cloth.

"This book thoroughly deserves its title of a 'Practical Treatise.' The whole process of gold milling, from the breaking of the quartz to the assay of the bullion, is described in clear and orderly narrative and with much, but not too much, fulness of detail."—*Saturday Review.*
"The work is a storehouse of information and valuable data, and we strongly recommend it to all professional men engaged in the gold-mining industry."—*Mining Journal.*

Silver, Metallurgy of.

THE METALLURGY OF SILVER: A Practical Treatise on the Amalgamation, Roasting and Lixiviation of Silver Ores, Including the Assaying, Melting and Refining of Silver Bullion. By M. EISSLER, Author of "The Metallurgy of Gold." Second Edition, Enlarged. With 150 Illustrations. Crown 8vo, 10s. 6d. cloth. [*Just published.*

"A practical treatise, and a technical work which we are convinced will supply a long-felt want amongst practical men, and at the same time be of value to students and others indirectly connected with the industries."—*Mining Journal.*
"From first to last the book is thoroughly sound and reliable."—*Colliery Guardian.*
"For chemists, practical miners, assayers and investors alike, we do not know of any work on the subject so handy and yet so comprehensive."—*Glasgow Herald.*

Silver-Lead, Metallurgy of.

THE METALLURGY OF ARGENTIFEROUS LEAD: A Practical Treatise on the Smelting of Silver-Lead Ores and the Refining of Lead Bullion. Including Reports on various Smelting Establishments and Descriptions of Modern Furnaces and Plants in Europe and America. By M. EISSLER, M.E., Author of "The Metallurgy of Gold," &c. Crown 8vo. 400 pp., with numerous Illustrations, 12s. 6d. cloth. [*Just published.*

Metalliferous Minerals and Mining.

TREATISE ON METALLIFEROUS MINERALS AND MINING. By D. C. DAVIES, F.G.S., Mining Engineer, &c., Author of "A Treatise on Slate and Slate Quarrying." Illustrated with numerous Wood Engravings. Fourth Edition, carefully Revised. Crown 8vo, 12s. 6d. cloth.

"Neither the practical miner nor the general reader interested in mines can have a better book for his companion and his guide."—*Mining Journal.* [*Mining World.*
"We are doing our readers a service in calling their attention to this valuable work."—
"As a history of the present state of mining throughout the world this book has a real value, and it supplies an actual want."—*Athenæum.*

Earthy Minerals and Mining.

A TREATISE ON EARTHY & OTHER MINERALS AND MINING. By D. C. DAVIES, F.G.S. Uniform with, and forming a Companion Volume to, the same Author's "Metalliferous Minerals and Mining." With 76 Wood Engravings. Second Edition. Crown 8vo, 12s. 6d. cloth.

"We do not remember to have met with any English work on mining matters that contains the same amount of information packed in equally convenient form."—*Academy.*
"We should be inclined to rank it as among the very best of the handy technical and trades manuals which have recently appeared."—*British Quarterly Review.*

Mineral Surveying and Valuing.

THE MINERAL SURVEYOR AND VALUER'S COMPLETE GUIDE, comprising a Treatise on Improved Mining Surveying and the Valuation of Mining Properties, with New Traverse Tables. By WM. LINTERN, Mining and Civil Engineer. Third Edition, with an Appendix on "Magnetic and Angular Surveying," with Records of the Peculiarities of Needle Disturbances. With Four Plates of Diagrams, Plans, &c. 12mo, 4s. cloth.

"Mr. Lintern's book forms a valuable and thoroughly trustworthy guide."—*Iron and Coal Trades Review.*

"This new edition must be of the highest value to colliery surveyors, proprietors and managers."—*Colliery Guardian.*

Asbestos and its Uses.

ASBESTOS: Its Properties, Occurrence and Uses. With some Account of the Mines of Italy and Canada. By ROBERT H. JONES. With Eight Collotype Plates and other Illustrations. Crown 8vo, 12s. 6d. cloth.

"An interesting and invaluable work."—*Colliery Guardian.*

"We counsel our readers to get this exceedingly interesting work for themselves; they will find in it much that is suggestive, and a great deal that is of immediate and practical usefulness."—*Builder.*

"A valuable addition to the architect's and engineer's library."—*Building News.*

Underground Pumping Machinery.

MINE DRAINAGE. Being a Complete and Practical Treatise on Direct-Acting Underground Steam Pumping Machinery, with a Description of a large number of the best known Engines, their General Utility and the Special Sphere of their Action, the Mode of their Application, and their merits compared with other forms of Pumping Machinery. By STEPHEN MICHELL. 8vo, 15s. cloth.

"Will be highly esteemed by colliery owners and lessees, mining engineers, and students generally who require to be acquainted with the best means of securing the drainage of mines. It is a most valuable work, and stands almost alone in the literature of steam pumping machinery."—*Colliery Guardian.*

"Much valuable information is given, so that the book is thoroughly worthy of an extensive circulation amongst practical men and purchasers of machinery."—*Mining Journal.*

Mining Tools.

A MANUAL OF MINING TOOLS. For the Use of Mine Managers, Agents, Students, &c. By WILLIAM MORGANS, Lecturer on Practical Mining at the Bristol School of Mines. 12mo, 2s. 6d. cloth limp.

ATLAS OF ENGRAVINGS to Illustrate the above, containing 235 Illustrations of Mining Tools, drawn to scale. 4to, 4s. 6d. cloth.

"Students in the science of mining, and overmen, captains, managers and viewers may gain practical knowledge and useful hints by the study of Mr. Morgans' manual."—*Colliery Guardian.*

"A valuable work, which will tend materially to improve our mining literature."—*Mining Journal.*

Coal Mining.

COAL AND COAL MINING: A Rudimentary Treatise on. By the late Sir WARINGTON W. SMYTH, M.A., F.R.S., &c., Chief Inspector of the Mines of the Crown. Seventh Edition, Revised and Enlarged. With numerous Illustrations. 12mo, 4s. cloth boards.

"As an outline is given of every known coal-field in this and other countries, as well as of the principal methods of working, the book will doubtless interest a very large number of readers."—*Mining Journal.*

Subterraneous Surveying.

SUBTERRANEOUS SURVEYING, Elementary and Practical Treatise on, with and without the Magnetic Needle. By THOMAS FENWICK, Surveyor of Mines, and THOMAS BAKER, C.E. Illust. 12mo, 3s. cloth boards.

Granite Quarrying.

GRANITES AND OUR GRANITE INDUSTRIES. By GEORGE F. HARRIS, F.G.S., Membre de la Société Belge de Géologie, Lecturer on Economic Geology at the Birkbeck Institution, &c. With Illustrations. Crown 8vo, 2s. 6d. cloth.

"A clearly and well-written manual for persons engaged or interested in the granite industry."—*Scotsman.*

"An interesting work, which will be deservedly esteemed."—*Colliery Guardian.*

"An exceedingly interesting and valuable monograph on a subject which has hitherto received unaccountably little attention in the shape of systematic literary treatment."—*Scottish Leader.*

ELECTRICITY ELECTRICAL ENGINEERING, etc.

Electrical Engineering.

THE ELECTRICAL ENGINEER'S POCKET-BOOK OF MODERN RULES, FORMULÆ, TABLES AND DATA. By H. R. KEMPE, M.Inst.E.E., A.M.Inst C.E., Technical Officer Postal Telegraphs, Author of "A Handbook of Electrical Testing," &c. With numerous Illustrations, royal 32mo, oblong, 5s. leather. [*Just published.*

"There is very little in the shape of formulæ or data which the electrician is likely to want in a hurry which cannot be found in its pages."—*Practical Engineer.*
"A very useful book of reference for daily use in practical electrical engineering and its various applications to the industries of the present day."—*Iron.*
"It is the best book of its kind."—*Electrical Engineer.*
"The Electrical Engineer's Pocket-Book is a good one."—*Electrician.*
"Strongly recommended to those engaged in the various electrical industries."—*Electrical Review.*

Electric Lighting.

ELECTRIC LIGHT FITTING: A Handbook for Working Electrical Engineers, embodying Practical Notes on Installation Management. By JOHN W. URQUHART, Electrician, Author of "Electric Light," &c. With numerous Illustrations, crown 8vo, 5s. cloth. [*Just published.*

"This volume deals with what may be termed the mechanics of electric lighting, and is addressed to men who are already engaged in the work or are training for it. The work traverses a great deal of ground, and may be read as a sequel to the same author's useful work on 'Electric Light.'"—*Electrician.*
"This is an attempt to state in the simplest language the precautions which should be adopted in instaling the electric light, and to give information for the guidance of those who have to run the plant when installed. The book is well worth the perusal of the workmen for whom it is written."—*Electrical Review.*
"Eminently practical and useful. . . . Ought to be in the hands of everyone in charge of an electric light plant."—*Electrical Engineer.*
"A really capital book, which we have no hesitation in recommending to the notice of working electricians and electrical engineers."—*Mechanical World.*

Electric Light.

ELECTRIC LIGHT: Its Production and Use. Embodying Plain Directions for the Treatment of Dynamo-Electric Machines, Batteries, Accumulators, and Electric Lamps. By J. W. URQUHART, C.E., Author of "Electric Light Fitting," &c. Fourth Edition, Revised, with Large Additions and 145 Illustrations. Crown 8vo, 7s. 6d. cloth. [*Just published.*

"The book is by far the best that we have yet met with on the subject."—*Athenæum.*
"It is the only work at present available which gives, in language intelligible for the most part to the ordinary reader, a general but concise history of the means which have been adopted up to the present time in producing the electric light."—*Metropolitan.*
"The book contains a general account of the means adopted in producing the electric light, not only as obtained from voltaic or galvanic batteries, but treats at length of the dynamo-electric machine in several of its forms."—*Colliery Guardian.*

Construction of Dynamos.

DYNAMO CONSTRUCTION: A Practical Handbook for the Use of Engineer Constructors and Electricians in Charge. With Examples of leading English, American and Continental Dynamos and Motors. By J. W. URQUHART, Author of "Electric Light," &c. Crown 8vo, 7s. 6d. cloth. [*Just published.*

"The author has produced a book for which a demand has long existed. The subject is treated in a thoroughly practical manner."—*Mechanical World.*

Dynamic Electricity and Magnetism.

THE ELEMENTS OF DYNAMIC ELECTRICITY AND MAGNETISM. By PHILIP ATKINSON, A.M., Ph.D. Crown 8vo. 400 pp. With 120 Illustrations. 10s. 6d. cloth. [*Just published.*

Text Book of Electricity.

THE STUDENT'S TEXT-BOOK OF ELECTRICITY. By HENRY M. NOAD, Ph.D., F.R.S., F.C.S. New Edition, carefully Revised. With an Introduction and Additional Chapters, by W. H. PREECE, M.I.C.E., Vice-President of the Society of Telegraph Engineers, &c. With 470 Illustrations. Crown 8vo, 12s. 6d. cloth.

"We can recommend Dr. Noad's book for clear style, great range of subject, a good index and a plethora of woodcuts. Such collections as the present are indispensable."—*Athenæum.*
"An admirable text book for every student — beginner or advanced — of electricity."—*Engineering.*

Electric Lighting.

THE ELEMENTARY PRINCIPLES OF ELECTRIC LIGHTING. By ALAN A. CAMPBELL SWINTON, Associate I.E.E. Second Edition, Enlarged and Revised. With 16 Illustrations. Crown 8vo, 1s. 6d. cloth.

"Anyone who desires a short and thoroughly clear exposition of the elementary principles of electric-lighting cannot do better than read this little work."—*Bradford Observer.*

Electricity.

A MANUAL OF ELECTRICITY: *Including Galvanism, Magnetism, Dia-Magnetism, Electro-Dynamics, Magno-Electricity, and the Electric Telegraph.* By HENRY M. NOAD, Ph.D., F.R.S., F.C.S. Fourth Edition. With 500 Woodcuts. 8vo, £1 4s. cloth.

"It is worthy of a place in the library of every public institution."—*Mining Journal.*

Dynamo Construction.

HOW TO MAKE A DYNAMO: *A Practical Treatise for Amateurs.* Containing numerous Illustrations and Detailed Instructions for Constructing a Small Dynamo, to Produce the Electric Light. By ALFRED CROFTS. Third Edition, Revised and Enlarged. Crown 8vo, 2s. cloth.

"The instructions given in this unpretentious little book are sufficiently clear and explicit to enable any amateur mechanic possessed of average skill and the usual tools to be found in an amateur's workshop, to build a practical dynamo machine."—*Electrician.*

NATURAL SCIENCE, etc.

Pneumatics and Acoustics.

PNEUMATICS: *including Acoustics and the Phenomena of Wind Currents,* for the Use of Beginners. By CHARLES TOMLINSON, F.R.S. F.C.S., &c. Fourth Edition, Enlarged. 12mo, 1s. 6d. cloth.

"Beginners in the study of this important application of science could not have a better manual."—*Scotsman.* "A valuable and suitable text-book for students of Acoustics and the Phenomena of Wind Currents."—*Schoolmaster.*

Conchology.

A MANUAL OF THE MOLLUSCA: *Being a Treatise on Recent and Fossil Shells.* By S. P. WOODWARD, A.L.S., F.G.S., late Assistant Palæontologist in the British Museum. With an Appendix on *Recent and Fossil Conchological Discoveries,* by RALPH TATE, A.L.S., F.G.S. Illustrated by A. N. WATERHOUSE and JOSEPH WILSON LOWRY. With 23 Plates and upwards of 300 Woodcuts. Reprint of Fourth Ed., 1880. Cr. 8vo, 7s. 6d. cl.

"A most valuable storehouse of conchological and geological information."—*Science Gossip.*

Geology.

RUDIMENTARY TREATISE ON GEOLOGY, PHYSICAL AND HISTORICAL. Consisting of "Physical Geology," which sets forth the leading Principles of the Science; and "Historical Geology," which treats of the Mineral and Organic Conditions of the Earth at each successive epoch, especial reference being made to the British Series of Rocks. By RALPH TATE, A.L.S., F.G.S., &c. With 250 Illustrations. 12mo, 5s. cloth.

"The fulness of the matter has elevated the book into a manual. Its information is exhaustive and well arranged."—*School Board Chronicle.*

Geology and Genesis.

THE TWIN RECORDS OF CREATION; *or, Geology and Genesis: their Perfect Harmony and Wonderful Concord.* By GEORGE W. VICTOR LE VAUX. Numerous Illustrations. Fcap. 8vo, 5s. cloth.

"A valuable contribution to the evidences of Revelation, and disposes very conclusively of the arguments of those who would set God's Works against God's Word."—*The Rock.*

The Constellations.

STAR GROUPS: *A Student's Guide to the Constellations.* By J. ELLARD GORE, F.R.A.S., M.R.I.A., &c., Author of "The Scenery of the Heavens." With 30 Maps. Small 4to, 5s. cloth, silvered. [*Just published.*

Astronomy.

ASTRONOMY. By the late Rev. ROBERT MAIN, M.A., F.R.S., formerly Radcliffe Observer at Oxford. Third Edition, Revised and Corrected to the present time, by W. T. LYNN, B.A., F.R.A.S. 12mo, 2s. cloth.

"A sound and simple treatise, very carefully edited, and a capital book for beginners."—*Knowledge.* [*tional Times.*

"Accurately brought down to the requirements of the present time by Mr. Lynn."—*Educa-*

DR. LARDNER'S COURSE OF NATURAL PHILOSOPHY.

THE HANDBOOK OF MECHANICS. Enlarged and almost re-written by BENJAMIN LOEWY, F.R.A.S. With 378 Illustrations. Post 8vo, 6s. cloth.

"The perspicuity of the original has been retained, and chapters which had become obsolete have been replaced by others of more modern character. The explanations throughout are studiously popular, and care has been taken to show the application of the various branches of physics to the industrial arts, and to the practical business of life."—*Mining Journal.*

"Mr. Loewy has carefully revised the book, and brought it up to modern requirements."—*Nature.*

"Natural philosophy has had few exponents more able or better skilled in the art of popularising the subject than Dr. Lardner; and Mr. Loewy is doing good service in fitting this treatise, and the others of the series, for use at the present time."—*Scotsman.*

THE HANDBOOK OF HYDROSTATICS AND PNEUMATICS. New Edition, Revised and Enlarged, by BENJAMIN LOEWY, F.R.A.S. With 236 Illustrations. Post 8vo, 5s. cloth.

"For those 'who desire to attain an accurate knowledge of physical science without the profound methods of mathematical investigation,' this work is not merely intended, but well adapted."—*Chemical News.*

"The volume before us has been carefully edited, augmented to nearly twice the bulk of the former edition, and all the most recent matter has been added. . . . It is a valuable text-book."—*Nature.*

"Candidates for pass examinations will find it, we think, specially suited to their requirements."—*English Mechanic.*

THE HANDBOOK OF HEAT. Edited and almost entirely re-written by BENJAMIN LOEWY, F.R.A.S., &c. 117 Illustrations. Post 8vo, 6s. cloth.

"The style is always clear and precise, and conveys instruction without leaving any cloudiness or lurking doubts behind."—*Engineering.*

"A most exhaustive book on the subject on which it treats, and is so arranged that it can be understood by all who desire to attain an accurate knowledge of physical science. . . . Mr. Loewy has included all the latest discoveries in the varied laws and effects of heat."—*Standard.*

"A complete and handy text-book for the use of students and general readers."—*English Mechanic.*

THE HANDBOOK OF OPTICS. By DIONYSIUS LARDNER, D.C.L., formerly Professor of Natural Philosophy and Astronomy in University College, London. New Edition. Edited by T. OLVER HARDING, B.A. Lond., of University College, London. With 298 Illustrations. Small 8vo, 448 pages, 5s. cloth.

"Written by one of the ablest English scientific writers, beautifully and elaborately illustrated."—*Mechanic's Magazine.*

THE HANDBOOK OF ELECTRICITY, MAGNETISM, AND ACOUSTICS. By Dr. LARDNER. Ninth Thousand. Edit. by GEORGE CAREY FOSTER, B.A., F.C.S. With 400 Illustrations. Small 8vo, 5s. cloth.

"The book could not have been entrusted to anyone better calculated to preserve the terse and lucid style of Lardner, while correcting his errors and bringing up his work to the present state of scientific knowledge."—*Popular Science Review.*

THE HANDBOOK OF ASTRONOMY. Forming a Companion to the "Handbook of Natural Philosophy." By DIONYSIUS LARDNER, D.C.L., formerly Professor of Natural Philosophy and Astronomy in University College, London. Fourth Edition. Revised and Edited by EDWIN DUNKIN, F.R.A.S., Royal Observatory, Greenwich. With 38 Plates and upwards of 100 Woodcuts. In One Vol., small 8vo, 550 pages, 9s. 6d. cloth.

"Probably no other book contains the same amount of information in so compendious and well-arranged a form—certainly none at the price at which this is offered to the public."—*Athenæum.*

"We can do no other than pronounce this work a most valuable manual of astronomy, and we strongly recommend it to all who wish to acquire a general—but at the same time correct—acquaintance with this sublime science."—*Quarterly Journal of Science.*

"One of the most deservedly popular books on the subject . . . We would recommend not only the student of the elementary principles of the science, but he who aims at mastering the higher and mathematical branches of astronomy, not to be without this work beside him."—*Practical Magazine.*

Dr. Lardner's Electric Telegraph.

THE ELECTRIC TELEGRAPH. By Dr. LARDNER. Re-vised and Re-written by E. B. BRIGHT, F.R.A.S. 140 Illustrations. Small 8vo, 2s. 6d. cloth.

"One of the most readable books extant on the Electric Telegraph."—*English Mechanic.*

DR. LARDNER'S MUSEUM OF SCIENCE AND ART.

THE MUSEUM OF SCIENCE AND ART. Edited by DIONYSIUS LARDNER, D.C.L., formerly Professor of Natural Philosophy and Astronomy in University College, London. With upwards of 1,200 Engravings on Wood. In 6 Double Volumes, £1 1s., in a new and elegant cloth binding; or handsomely bound in half-morocco, 31s. 6d.

* OPINIONS OF THE PRESS.

"This series, besides affording popular but sound instruction on scientific subjects, with which the humblest man in the country ought to be acquainted, also undertakes that teaching of 'Common Things' which every well-wisher of his kind is anxious to promote. Many thousand copies of this serviceable publication have been printed, in the belief and hope that the desire for instruction and improvement widely prevails; and we have no fear that such enlightened faith will meet with disappointment."—*Times.*

"A cheap and interesting publication, alike informing and attractive. The papers combine subjects of importance and great scientific knowledge, considerable inductive powers, and a popular style of treatment."—*Spectator.*

"The 'Museum of Science and Art' is the most valuable contribution that has ever been made to the Scientific Instruction of every class of society."—Sir DAVID BREWSTER, in the *North British Review.*

"Whether we consider the liberality and beauty of the illustrations, the charm of the writing, or the durable interest of the matter, we must express our belief that there is hardly to be found among the new books one that would be welcomed by people of so many ages and classes as a valuable present."—*Examiner.*

* *Separate books formed from the above, suitable for Workmen's Libraries, Science Classes, etc.*

Common Things Explained. Containing Air, Earth, Fire, Water, Time, Man, the Eye, Locomotion, Colour, Clocks and Watches, &c. 233 Illustrations, cloth gilt, 5s.

The Microscope. Containing Optical Images, Magnifying Glasses, Origin and Description of the Microscope, Microscopic Objects, the Solar Microscope, Microscopic Drawing and Engraving, &c. 147 Illustrations, cloth gilt, 2s.

Popular Geology. Containing Earthquakes and Volcanoes, the Crust of the Earth, &c. 201 Illustrations, cloth gilt, 2s. 6d.

Popular Physics. Containing Magnitude and Minuteness, the Atmosphere, Meteoric Stones, Popular Fallacies, Weather Prognostics, the Thermometer, the Barometer, Sound, &c. 85 Illustrations, cloth gilt, 2s. 6d.

Steam and its Uses. Including the Steam Engine, the Locomotive, and Steam Navigation. 89 Illustrations, cloth gilt, 2s.

Popular Astronomy. Containing How to observe the Heavens—The Earth, Sun, Moon, Planets, Light, Comets, Eclipses, Astronomical Influences, &c. 182 Illustrations, 4s. 6d.

The Bee and White Ants: Their Manners and Habits. With Illustrations of Animal Instinct and Intelligence. 135 Illustrations, cloth gilt, 2s.

The Electric Telegraph Popularized. To render intelligible to all who can Read, irrespective of any previous Scientific Acquirements, the various forms of Telegraphy in Actual Operation. 100 Illustrations, cloth gilt, 1s. 6d.

Dr. Lardner's School Handbooks.

NATURAL PHILOSOPHY FOR SCHOOLS. By Dr. LARDNER. 328 Illustrations. Sixth Edition. One Vol., 3s. 6d. cloth.

"A very convenient class-book for junior students in private schools. It is intended to convey, in clear and precise terms, general notions of all the principal divisions of Physical Science."—*British Quarterly Review.*

ANIMAL PHYSIOLOGY FOR SCHOOLS. By Dr. LARDNER. With 190 Illustrations. Second Edition. One Vol., 3s. 6d. cloth.

"Clearly written, well arranged, and excellently illustrated."—*Gardener's Chronicle.*

COUNTING-HOUSE WORK, TABLES, etc.

Introduction to Business.

LESSONS IN COMMERCE. By Professor R. GAMBARO, of the Royal High Commercial School at Genoa. Edited and Revised by JAMES GAULT, Professor of Commerce and Commercial Law in King's College, London. Crown 8vo, price about 3s. 6d. [*In the press.*

Accounts for Manufacturers.

FACTORY ACCOUNTS: Their Principles and Practice. A Handbook for Accountants and Manufacturers, with Appendices on the Nomenclature of Machine Details; the Income Tax Acts; the Rating of Factories; Fire and Boiler Insurance; the Factory and Workshop Acts, &c., including also a Glossary of Terms and a large number of Specimen Rulings. By EMILE GARCKE and J. M. FELLS. Third Edition. Demy 8vo, 250 pages, price 6s. strongly bound.

"A very interesting description of the requirements of Factory Accounts. . . . the principle of assimilating the Factory Accounts to the general commercial books is one which we thoroughly agree with."—*Accountants' Journal.*

"There are few owners of Factories who would not derive great benefit from the perusal of this most admirable work."—*Local Government Chronicle.*

Foreign Commercial Correspondence.

THE FOREIGN COMMERCIAL CORRESPONDENT: Being Aids to Commercial Correspondence in Five Languages—English, French, German, Italian and Spanish. By CONRAD E. BAKER. Second Edition, Revised. Crown 8vo, 3s. 6d. cloth.

"Whoever wishes to correspond in all the languages mentioned by Mr. Baker cannot do better than study this work, the materials of which are excellent and conveniently arranged."—*Athenæum.*

"A careful examination has convinced us that it is unusually complete, well arranged and reliable. The book is a thoroughly good one."—*Schoolmaster.*

Intuitive Calculations.

THE COMPENDIOUS CALCULATOR; or, Easy and Concise Methods of Performing the various Arithmetical Operations required in Commercial and Business Transactions, together with Useful Tables. By D. O'GORMAN. Corrected by Professor J. R. YOUNG. Twenty-seventh Ed., Revised by C. NORRIS. Fcap. 8vo, 2s. 6d. cloth; or, 3s. 6d. half-bound.

"It would be difficult to exaggerate the usefulness of a book like this to everyone engaged in commerce or manufacturing industry."—*Knowledge.*

"Supplies special and rapid methods for all kinds of calculations. Of great utility to persons engaged in any kind of commercial transactions."—*Scotsman.*

Modern Metrical Units and Systems.

MODERN METROLOGY: A Manual of the Metrical Units and Systems of the Present Century. With an Appendix containing a proposed English System. By LOWIS D'A. JACKSON, A.M.Inst.C.E., Author of "Aid to Survey Practice," &c. Large crown 8vo, 12s. 6d. cloth.

"The author has brought together much valuable and interesting information. . . . We cannot but recommend the work."—*Nature.*

"For exhaustive tables of equivalent weights and measures of all sorts, and for clear demonstrations of the effects of the various systems that have been proposed or adopted, Mr. Jackson's treatise is without a rival."—*Academy.*

The Metric System and the British Standards.

A SERIES OF METRIC TABLES, in which the British Standard Measures and Weights are compared with those of the Metric System at present in Use on the Continent. By C. H. DOWLING, C.E. 8vo, 10s. 6d. strongly bound.

"Their accuracy has been certified by Professor Airy, the Astronomer-Royal."—*Builder.*

"Mr. Dowling's Tables are well put together as a ready-reckoner for the conversion of one system into the other."—*Athenæum.*

Iron and Metal Trades' Calculator.

THE IRON AND METAL TRADES' COMPANION. For expeditiously ascertaining the Value of any Goods bought or sold by Weight, from 1s. per cwt. to 112s. per cwt., and from one farthing per pound to one shilling per pound. Each Table extends from one pound to 100 tons. To which are appended Rules on Decimals, Square and Cube Root, Mensuration of Superficies and Solids, &c.; also Tables of Weights of Materials, and other Useful Memoranda. By THOS. DOWNIE. Strongly bound in leather, 396 pp., 9s.

"A most useful set of tables. . . . Nothing like them before existed."—*Building News.*

"Although specially adapted to the iron and metal trades, the tables will be found useful in every other business in which merchandise is bought and sold by weight."—*Railway News.*

Calculator for Numbers and Weights Combined.

THE NUMBER, WEIGHT AND FRACTIONAL CALCU-LATOR. Containing upwards of 250,000 Separate Calculations, showing at a glance the value at 422 different rates, ranging from $\frac{1}{16}$th of a Penny to 20s. each, or per cwt., and £20 per ton, of any number of articles consecutively, from 1 to 470.—Any number of cwts., qrs., and lbs., from 1 cwt. to 470 cwts.—Any number of tons, cwts., qrs., and lbs., from 1 to 1,000 tons. By WILLIAM CHADWICK, Public Accountant. Third Edition, Revised and Improved. 8vo, price 18s., strongly bound for Office wear and tear.

*** *This work is specially adapted for the Apportionment of Mileage Charges for Railway Traffic.*

☞ *This comprehensive and entirely unique and original Calculator is adapted for the use of Accountants and Auditors, Railway Companies, Canal Companies, Shippers, Shipping Agents, General Carriers, etc.*

Ironfounders, Brassfounders, Metal Merchants, Iron Manufacturers, Ironmongers, Engineers, Machinists, Boiler Makers, Millwrights, Roofing, Bridge and Girder Makers, Colliery Proprietors, etc.

Timber Merchants, Builders, Contractors, Architects, Surveyors, Auctioneers, Valuers, Brokers, Mill Owners and Manufacturers, Mill Furnishers, Merchants and General Wholesale Tradesmen.

*** OPINIONS OF THE PRESS.

"The book contains the answers to questions, and not simply a set of ingenious puzzle methods of arriving at results. It is as easy of reference for any answer or any number of answers as a dictionary, and the references are even more quickly made. For making up accounts or estimates, the book must prove invaluable to all who have any considerable quantity of calculations involving price and measure in any combination to do."—*Engineer.*

"The most perfect work of the kind yet prepared."—*Glasgow Herald.*

Comprehensive Weight Calculator.

THE WEIGHT CALCULATOR. Being a Series of Tables upon a New and Comprehensive Plan, exhibiting at One Reference the exact Value of any Weight from 1 lb. to 15 tons, at 300 Progressive Rates, from 1d. to 168s. per cwt., and containing 186,000 Direct Answers, which, with their Combinations, consisting of a single addition (mostly to be performed at sight), will afford an aggregate of 10,266,000 Answers; the whole being calculated and designed to ensure correctness and promote despatch. By HENRY HARBEN, Accountant. Fourth Edition, carefully Corrected. Royal 8vo, strongly half-bound, £1 5s.

"A practical and useful work of reference for men of business generally; it is the best of the kind we have seen."—*Ironmonger.*

"Of priceless value to business men. It is a necessary book in all mercantile offices."—*Sheffield Independent.*

Comprehensive Discount Guide.

THE DISCOUNT GUIDE. Comprising several Series of Tables for the use of Merchants, Manufacturers, Ironmongers, and others, by which may be ascertained the exact Profit arising from any mode of using Discounts, either in the Purchase or Sale of Goods, and the method of either Altering a Rate of Discount or Advancing a Price, so as to produce, by one operation, a sum that will realise any required profit after allowing one or more Discounts: to which are added Tables of Profit or Advance from 1¼ to 90 per cent., Tables of Discount from 1¼ to 98¾ per cent., and Tables of Commission, &c., from ⅛ to 10 per cent. By HENRY HARBEN, Accountant, Author of "The Weight Calculator." New Edition, carefully Revised and Corrected. Demy 8vo, 544 pp. half-bound, £1 5s.

"A book such as this can only be appreciated by business men, to whom the saving of time means saving of money. We have the high authority of Professor J. R. Young that the tables throughout the work are constructed upon strictly accurate principles. The work is a mode of typographical clearness, and must prove of great value to merchants, manufacturers, and general traders."—*British Trade Journal.*

Iron Shipbuilders' and Merchants' Weight Tables.

IRON-PLATE WEIGHT TABLES: For Iron Shipbuilders, Engineers and Iron Merchants. Containing the Calculated Weights of upwards of 150,000 different sizes of Iron Plates, from 1 foot by 6 in. by ¼ in. to 10 feet by 5 feet by 1 in. Worked out on the basis of 40 lbs. to the square foot of Iron of 1 inch in thickness. Carefully compiled and thoroughly Revised by H. BURLINSON and W. H. SIMPSON. Oblong 4to, 25s. half-bound.

"This work will be found of great utility. The authors have had much practical experience of what is wanting in making estimates; and the use of the book will save much time in making elaborate calculations."—*English Mechanic.*

INDUSTRIAL AND USEFUL ARTS.

Soap-making.
THE ART OF SOAP-MAKING: *A Practical Handbook of the Manufacture of Hard and Soft Soaps, Toilet Soaps, etc.* Including many New Processes, and a Chapter on the Recovery of Glycerine from Waste Leys. By ALEXANDER WATT, Author of "Electro-Metallurgy Practically Treated," &c. With numerous Illustrations. Fourth Edition, Revised and Enlarged. Crown 8vo, 7s. 6d. cloth.

"The work will prove very useful, not merely to the technological student, but o h? practical soap-boiler who wishes to understand the theory of his art."—*Chemical News.*

"Mr. Watt's book is a thoroughly practical treatise on an art which has almost no literature in our language. We congratulate the author on the success of his endeavour to fill a void in English technical literature."—*Nature.*

Paper Making.
THE ART OF PAPER MAKING: *A Practical Handbook of the Manufacture of Paper from Rags, Esparto, Straw and other Fibrous Materials,* Including the Manufacture of Pulp from Wood Fibre, with a Description of the Machinery and Appliances used. To which are added Details of Processes for Recovering Soda from Waste Liquors. By ALEXANDER WATT. With Illustrations. Crown 8vo, 7s. 6d. cloth.

"This book is succinct, lucid, thoroughly practical, and includes everything of interest to the modern paper maker. It is the latest, most practical and most complete work on the paper-making art before the British public."—*Paper Record.*

"It may be regarded as the standard work on the subject. The book is full of valuable information. The 'Art of Paper-making,' is in every respect a model of a text-book, either for a technical class or for the private student."—*Paper and Printing Trades Journal.*

"Admirably adapted for general as well as ordinary technical reference, and as a handbook for students in technical education may be warmly commended."—*The Paper Maker's Monthly Journal.*

Leather Manufacture.
THE ART OF LEATHER MANUFACTURE. Being a Practical Handbook, in which the Operations of Tanning, Currying, and Leather Dressing are fully Described, the Principles of Tanning Explained and many Recent Processes introduced. By ALEXANDER WATT, Author of "Soap-Making," &c. With numerous Illustrations. Second Edition. Crown 8vo, 9s. cloth.

"A sound, comprehensive treatise on tanning and its accessories. This book is an eminently valuable production, which redounds to the credit of both author and publishers."—*Chemical Review.*

"This volume is technical without being tedious, comprehensive and complete without being prosy, and it bears on every page the impress of a master hand. We have never come across a better trade treatise, nor one that so thoroughly supplied an absolute want."—*Shoe and Leather Trades' Chronicle.*

Boot and Shoe Making.
THE ART OF BOOT AND SHOE-MAKING. A Practical Handbook, including Measurement, Last-Fitting, Cutting-Out, Closing and Making, with a Description of the most approved Machinery employed. By JOHN B. LENO, late Editor of *St. Crispin*, and *The Boot and Shoe-Maker*. With numerous Illustrations. Third Edition. 12mo, 2s. cloth limp.

"This excellent treatise is by far the best work ever written on the subject. A new work, embracing all modern improvements, was much wanted. This want is now satisfied. The chapter on clicking, which shows how waste may be prevented, will save fifty times the price of the book." —*Scottish Leather Trader.*

Dentistry.
MECHANICAL DENTISTRY: *A Practical Treatise on the Construction of the various kinds of Artificial Dentures.* Comprising also Useful Formulæ, Tables and Receipts for Gold Plate, Clasps, Solders, &c. &c. By CHARLES HUNTER. Third Edition, Revised. With upwards of 100 Wood Engravings. Crown 8vo, 3s. 6d. cloth.

"The work is very practical."—*Monthly Review of Dental Surgery.*

"We can strongly recommend Mr. Hunter's treatise to all students preparing for the profession of dentistry, as well as to every mechanical dentist."—*Dublin Journal of Medical Science.*

Wood Engraving.
WOOD ENGRAVING: *A Practical and Easy Introduction to the Study of the Art.* By WILLIAM NORMAN BROWN. Second Edition. With numerous Illustrations. 12mo, 1s. 6d. cloth limp.

"The book is clear and complete, and will be useful to anyone wanting to understand the first elements of the beautiful art of wood engraving."—*Graphic.*

HANDYBOOKS FOR HANDICRAFTS. By PAUL N. HASLUCK.

Metal Turning.
THE METAL TURNER'S HANDYBOOK. *A Practical Manual for Workers at the Foot-Lathe:* Embracing Information on the Tools, Appliances and Processes employed in Metal Turning. By PAUL N. HASLUCK, Author of "Lathe-Work." With upwards of One Hundred Illustrations. Second Edition, Revised. Crown 8vo, 2s. cloth.
"Clearly and concisely written, excellent in every way."—*Mechanical World.*

Wood Turning.
THE WOOD TURNER'S HANDYBOOK. *A Practical Manual for Workers at the Lathe:* Embracing Information on the Tools, Appliances and Processes Employed in Wood Turning. By PAUL N. HASLUCK. With upwards of One Hundred Illustrations. Crown 8vo, 2s. cloth.
"We recommend the book to young turners and amateurs. A multitude of workmen have hitherto sought in vain for a manual of this special industry."—*Mechanical World.*

WOOD AND METAL TURNING. By P. N. HASLUCK. (Being the Two preceding Vols. bound together.) 300 pp., with upwards of 200 Illustrations, crown 8vo, 3s. 6d. cloth.

Watch Repairing.
THE WATCH JOBBER'S HANDYBOOK. *A Practical Manual on Cleaning, Repairing and Adjusting.* Embracing Information on the Tools, Materials, Appliances and Processes Employed in Watchwork. By PAUL N. HASLUCK. With upwards of One Hundred Illustrations. Cr. 8vo, 2s. cloth.
"All young persons connected with the trade should acquire and study this excellent, and at the same time, inexpensive work."—*Clerkenwell Chronicle.*

Clock Repairing.
THE CLOCK JOBBER'S HANDYBOOK: *A Practical Manual on Cleaning, Repairing and Adjusting.* Embracing Information on the Tools, Materials, Appliances and Processes Employed in Clockwork. By PAUL N. HASLUCK. With upwards of 100 Illustrations. Cr. 8vo, 2s. cloth.
"Of inestimable service to those commencing the trade."—*Coventry Standard.*

WATCH AND CLOCK JOBBING. By P. N. HASLUCK. (Being the Two preceding Vols. bound together.) 320 pp., with upwards of 200 Illustrations, crown 8vo, 3s. 6d. cloth.

Pattern Making.
THE PATTERN MAKER'S HANDYBOOK. A Practical Manual, embracing Information on the Tools, Materials and Appliances employed in Constructing Patterns for Founders. By PAUL N. HASLUCK. With One Hundred Illustrations. Crown 8vo, 2s. cloth.
"This handy volume contains sound information of considerable value to students and artificers."—*Hardware Trades Journal.*

Mechanical Manipulation.
THE MECHANIC'S WORKSHOP HANDYBOOK. *A Practical Manual on Mechanical Manipulation.* Embracing Information on various Handicraft Processes, with Useful Notes and Miscellaneous Memoranda. By PAUL N. HASLUCK. Crown 8vo, 2s. cloth.
"It is a book which should be found in every workshop, as it is one which will be continually referred to for a very great amount of standard information."—*Saturday Review.*

Model Engineering.
THE MODEL ENGINEER'S HANDYBOOK: *A Practical Manual on Model Steam Engines.* Embracing Information on the Tools, Materials and Processes Employed in their Construction. By PAUL N. HASLUCK. With upwards of 100 Illustrations. Crown 8vo, 2s. cloth.
"By carefully going through the work, amateurs may pick up an excellent notion of the construction of full-sized steam engines."—*Telegraphic Journal.*

Cabinet Making.
THE CABINET WORKER'S HANDYBOOK: A Practical Manual, embracing Information on the Tools, Materials, Appliances and Processes employed in Cabinet Work. By PAUL N. HASLUCK, Author of "Lathe Work," &c. With upwards of 100 Illustrations. Crown 8vo, 2s. cloth. [*Glasgow Herald.*
"Thoroughly practical throughout. The amateur worker in wood will find it most useful."—

INDUSTRIAL AND USEFUL ARTS. 33

Electrolysis of Gold, Silver, Copper, etc.

ELECTRO-DEPOSITION: A Practical Treatise on the Electrolysis of Gold, Silver, Copper, Nickel, and other Metals and Alloys. With descriptions of Voltaic Batteries, Magneto and Dynamo-Electric Machines, Thermopiles, and of the Materials and Processes used in every Department of the Art, and several Chapters on Electro-Metallurgy. By ALEXANDER WATT. Third Edition, Revised and Corrected. Crown 8vo, 9s. cloth.

"Eminently a book for the practical worker in electro-deposition. It contains practical descriptions of methods, processes and materials as actually pursued and used in the workshop."—*Engineer.*

Electro-Metallurgy.

ELECTRO-METALLURGY; Practically Treated. By ALEXANDER WATT, Author of "Electro-Deposition," &c. Ninth Edition, Enlarged and Revised, with Additional Illustrations, and including the most recent Processes. 12mo, 4s. cloth boards.

"From this book both amateur and artisan may learn everything necessary for the successful prosecution of electroplating."—*Iron.*

Electroplating.

ELECTROPLATING: A Practical Handbook on the Deposition of Copper, Silver, Nickel, Gold, Aluminium, Brass, Platinum, &c. &c. With Descriptions of the Chemicals, Materials, Batteries and Dynamo Machines used in the Art. By J. W. URQUHART, C.E. Second Edition, with Additions. Numerous Illustrations. Crown 8vo, 5s. cloth.

"An excellent practical manual."—*Engineering.*
"An excellent work, giving the newest information."—*Horological Journal.*

Electrotyping.

ELECTROTYPING: The Reproduction and Multiplication of Printing Surfaces and Works of Art by the Electro-deposition of Metals. By J. W. URQUHART, C.E. Crown 8vo, 5s. cloth.

"The book is thoroughly practical. The reader is, therefore, conducted through the leading laws of electricity, then through the metals used by electrotypers, the apparatus, and the depositing processes, up to the final preparation of the work."—*Art Journal.*

Horology.

A TREATISE ON MODERN HOROLOGY, in Theory and Practice. Translated from the French of CLAUDIUS SAUNIER, by JULIEN TRIPPLIN, F.R.A.S., and EDWARD RIGG, M.A., Assayer in the Royal Mint. With 78 Woodcuts and 22 Coloured Plates. Second Edition. Royal 8vo, £2 2s. cloth; £2 10s. half-calf.

"There is no horological work in the English language at all to be compared to this production of M. Saunier's for clearness and completeness. It is alike good as a guide for the student and as a reference for the experienced horologist and skilled workman."—*Horological Journal.*

"The latest, the most complete, and the most reliable of those literary productions to which continental watchmakers are indebted for the mechanical superiority over their English brethren —in fact, the Book of Books, is M. Saunier's 'Treatise.'"—*Watchmaker, Jeweller and Silversmith.*

Watchmaking.

THE WATCHMAKER'S HANDBOOK. A Workshop Companion for those engaged in Watchmaking and the Allied Mechanical Arts. From the French of CLAUDIUS SAUNIER. Enlarged by JULIEN TRIPPLIN, F.R.A.S., and EDWARD RIGG, M.A., Assayer in the Royal Mint. Woodcuts and Copper Plates. Third Edition, Revised. Crown 8vo, 9s. cloth.

"Each part is truly a treatise in itself. The arrangement is good and the language is clear and concise. It is an admirable guide for the young watchmaker."—*Engineering.*

"It is impossible to speak too highly of its excellence. It fulfils every requirement in a handbook intended for the use of a workman."—*Watch and Clockmaker.*

"This book contains an immense number of practical details bearing on the daily occupation of a watchmaker."—*Watchmaker and Metalworker* (Chicago).

Goldsmiths' Work.

THE GOLDSMITH'S HANDBOOK. By GEORGE E. GEE, Jeweller, &c. Third Edition, considerably Enlarged. 12mo, 3s. 6d. cl. bds.

"A good, sound educator, and will be accepted as an authority."—*Horological Journal.*

Silversmiths' Work.

THE SILVERSMITH'S HANDBOOK. By GEORGE E. GEE, Jeweller, &c. Second Edition, Revised, with numerous Illustrations. 12mo, 3s. 6d. cloth boards.

"Workers in the trade will speedily discover its merits when they sit down to study it."—*English Mechanic.*

*** *The above two works together, strongly half-bound, price 7s.*

Bread and Biscuit Baking.

THE BREAD AND BISCUIT BAKER'S AND SUGAR-BOILER'S ASSISTANT. Including a large variety of Modern Recipes. With Remarks on the Art of Bread-making. By ROBERT WELLS, Practical Baker. Second Edition, with Additional Recipes. Crown 8vo, 2s. cloth.
"A large number of wrinkles for the ordinary cook, as well as the baker."—*Saturday Review.*

Confectionery.

THE PASTRYCOOK AND CONFECTIONER'S GUIDE. For Hotels, Restaurants and the Trade in general, adapted also for Family Use. By ROBERT WELLS, Author of "The Bread and Biscuit Baker's and Sugar Boiler's Assistant." Crown 8vo, 2s. cloth.
"We cannot speak too highly of this really excellent work. In these days of keen competition our readers cannot do better than purchase this book."—*Bakers' Times.*

Ornamental Confectionery.

ORNAMENTAL CONFECTIONERY: A Guide for Bakers, Confectioners and Pastrycooks; including a variety of Modern Recipes, and Remarks on Decorative and Coloured Work. With 129 Original Designs. By ROBERT WELLS. Crown 8vo, 5s. cloth.
"A valuable work, and should be in the hands of every baker and confectioner. The illustrative designs are alone worth treble the amount charged for the whole work."—*Bakers' Times.*

Flour Confectionery.

THE MODERN FLOUR CONFECTIONER. Wholesale and Retail. Containing a large Collection of Recipes for Cheap Cakes, Biscuits, &c. With Remarks on the Ingredients used in their Manufacture, &c. By R. WELLS, Author of "Ornamental Confectionery," "The Bread and Biscuit Baker," "The Pastrycook's Guide," &c. Crown 8vo, 2s. cloth.

Laundry Work.

LAUNDRY MANAGEMENT. A Handbook for Use in Private and Public Laundries, Including Descriptive Accounts of Modern Machinery and Appliances for Laundry Work. By the EDITOR of "The Laundry Journal." With numerous Illustrations. Crown 8vo, 2s. 6d. cloth.

CHEMICAL MANUFACTURES & COMMERCE.

New Manual of Engineering Chemistry.

ENGINEERING CHEMISTRY: A Practical Treatise for the Use of Analytical Chemists, Engineers, Iron Masters, Iron Founders, Students, and others. Comprising Methods of Analysis and Valuation of the Principal Materials used in Engineering Work, with numerous Analyses, Examples, and Suggestions. By H. JOSHUA PHILLIPS, F.I.C., F.C.S. Analytical and Consulting Chemist to the Great Eastern Railway. Crown 8vo 320 pp., with Illustrations, 10s. 6d. cloth. [*Just published*
"In this work the author has rendered no small service to a numerous body of practical men . . . The analytical methods may be pronounced most satisfactory, being as accurate as the despatch required of engineering chemists permits."—*Chemical News.*

Analysis and Valuation of Fuels.

FUELS: SOLID, LIQUID AND GASEOUS, Their Analysis and Valuation. For the Use of Chemists and Engineers. By H. J. PHILLIPS, F.C.S., Analytical and Consulting Chemist to the Great Eastern Railway. Crown 8vo, 3s. 6d. cloth.
"Ought to have its place in the laboratory of every metallurgical establishment, and wherever fuel is used on a large scale."—*Chemical News.*
"Cannot fail to be of wide interest, especially at the present time."—*Railway News.*

Alkali Trade, Manufacture of Sulphuric Acid, etc.

A MANUAL OF THE ALKALI TRADE, including the Manufacture of Sulphuric Acid, Sulphate of Soda, and Bleaching Powder. By JOHN LOMAS. 390 pages. With 232 Illustrations and Working Drawings. Second Edition. Royal 8vo, £1 10s. cloth.
"This book is written by a manufacturer for manufacturers. The working details of the most approved forms of apparatus are given, and these are accompanied by no less than 232 wood engravings, all of which may be used for the purposes of construction."—*Athenæum.*

The Blowpipe.

THE BLOWPIPE IN CHEMISTRY, MINERALOGY, AND GEOLOGY. Containing all known Methods of Anhydrous Analysis, Working Examples, and Instructions for Making Apparatus. By Lieut.-Col. W. A Ross, R.A. With 120 Illustrations. New Edition. Crown 8vo, 5s. cloth.

"The student who goes through the course of experimentation here laid down will gain a better insight into inorganic chemistry and mineralogy than if he had 'got up' any of the best text-books of the day, and passed any number of examinations in their contents."—*Chemical News*.

Commercial Chemical Analysis.

THE COMMERCIAL HANDBOOK OF CHEMICAL ANALYSIS; or, Practical Instructions for the determination of the Intrinsic or Commercial Value of Substances used in Manufactures, Trades, and the Arts. By A. NORMANDY. New Edition by H. M. NOAD, F.R.S. Cr. 8vo, 12s. 6d. cl.

"Essential to the analysts appointed under the new Act. The most recent results are given, and the work is well edited and carefully written."—*Nature*.

Brewing.

A HANDBOOK FOR YOUNG BREWERS. By HERBERT EDWARDS WRIGHT, B.A. New Edition, much Enlarged. [*In the press.*

Dye-Wares and Colours.

THE MANUAL OF COLOURS AND DYE-WARES: Their Properties, Applications, Valuation, Impurities, and Sophistications. For the use of Dyers, Printers, Drysalters, Brokers, &c. By J. W. SLATER. Second Edition, Revised and greatly Enlarged. Crown 8vo, 7s. 6d. cloth.

"A complete encyclopædia of the *materia tinctoria*. The information given respecting each article is full and precise, and the methods of determining the value of articles such as these, so liable to sophistication, are given with clearness, and are practical as well as valuable."—*Chemist and Druggist*.

"There is no other work which covers precisely the same ground. To students preparing for examinations in dyeing and printing it will prove exceedingly useful."—*Chemical News*.

Pigments.

THE ARTIST'S MANUAL OF PIGMENTS. Showing their Composition, Conditions of Permanency, Non-Permanency, and Adulterations; Effects in Combination with Each Other and with Vehicles; and the most Reliable Tests of Purity. By H. C. STANDAGE. Second Edition. Crown 8vo, 2s. 6d. cloth.

"This work is indeed *multum-in-parvo*, and we can, with good conscience, recommend it to all who come in contact with pigments, whether as makers, dealers or users."—*Chemical Review*.

Gauging. Tables and Rules for Revenue Officers, Brewers, etc.

A POCKET BOOK OF MENSURATION AND GAUGING: Containing Tables, Rules and Memoranda for Revenue Officers, Brewers, Spirit Merchants, &c. By J. B. MANT (Inland Revenue). Second Edition Revised. Oblong 18mo, 4s. leather, with elastic band.

"This handy and useful book is adapted to the requirements of the Inland Revenue Department, and will be a favourite book of reference."—*Civilian*.

"Should be in the hands of every practical brewer."—*Brewers' Journal*.

AGRICULTURE, FARMING, GARDENING, etc.

Youatt and Burn's Complete Grazier.

THE COMPLETE GRAZIER, and FARMER'S and CATTLE-BREEDER'S ASSISTANT. Including the Breeding, Rearing, and Feeding of Stock; Management of the Dairy, Culture and Management of Grass Land, and of Grain and Root Crops, &c. By W. YOUATT and R. SCOTT BURN. An entirely New Edition, partly Re-written and greatly Enlarged, by W. FREAM, B.Sc.Lond., LL.D. In medium 8vo, about 1,000 pp. [*In the press.*

Agricultural Facts and Figures.

NOTE-BOOK OF AGRICULTURAL FACTS AND FIGURES FOR FARMERS AND FARM STUDENTS. By PRIMROSE MCCONNELL, late Professor of Agriculture, Glasgow Veterinary College. Third Edition Royal 32mo, 4s. leather.

"The most complete and comprehensive Note-book for Farmers and Farm Students that we have seen. It literally teems with information, and we can cordially recommend it to all connected with agriculture."—*North British Agriculturist*.

Flour Manufacture, Milling, etc.

FLOUR MANUFACTURE: A Treatise on Milling Science and Practice. By FRIEDRICH KICK, Imperial Regierungsrath, Professor of Mechanical Technology in the Imperial German Polytechnic Institute, Prague. Translated from the Second Enlarged and Revised Edition with Supplement. By H. H. P. POWLES, A.M.I.C.E. Nearly 400 pp. Illustrated with 28 Folding Plates, and 167 Woodcuts. Royal 8vo, 25s. cloth.

"This valuable work is, and will remain, the standard authority on the science of milling. The miller who has read and digested this work will have laid the foundation, so to speak, of a successful career; he will have acquired a number of general principles which he can proceed to apply. In this handsome volume we at last have the accepted text-book of modern milling in good, sound English, which has little, if any, trace of the German idiom."—*The Miller.*

"The appearance of this celebrated work in English is very opportune, and British millers will, we are sure, not be slow in availing themselves of its pages."—*Millers' Gazette.*

Small Farming.

SYSTEMATIC SMALL FARMING; or, The Lessons of my Farm. Being an Introduction to Modern Farm Practice for Small Farmers in the Culture of Crops; The Feeding of Cattle; The Management of the Dairy, Poultry and Pigs, &c. &c. By ROBERT SCOTT BURN, Author of "Outlines of Landed Estates' Management." Numerous Illusts., cr. 8vo, 6s. cloth.

"This is the completest book of its class we have seen, and one which every amateur farmer will read with pleasure and accept as a guide."—*Field.*

"The volume contains a vast amount of useful information. No branch of farming is left untouched, from the labour to be done to the results achieved. It may be safely recommended to all who think they will be in paradise when they buy or rent a three-acre farm."—*Glasgow Herald.*

Modern Farming.

OUTLINES OF MODERN FARMING. By R. SCOTT BURN. Soils, Manures, and Crops—Farming and Farming Economy—Cattle, Sheep, and Horses — Management of Dairy, Pigs and Poultry — Utilisation of Town-Sewage, Irrigation, &c. Sixth Edition. In One Vol., 1,250 pp., half-bound, profusely Illustrated, 12s.

"The aim of the author has been to make his work at once comprehensive and trustworthy, and in this aim he has succeeded to a degree which entitles him to much credit."—*Morning Advertiser.* "No farmer should be without this book."—*Banbury Guardian.*

Agricultural Engineering.

FARM ENGINEERING, THE COMPLETE TEXT-BOOK OF. Comprising Draining and Embanking; Irrigation and Water Supply; Farm Roads, Fences, and Gates; Farm Buildings, their Arrangement and Construction, with Plans and Estimates; Barn Implements and Machines; Field Implements and Machines; Agricultural Surveying, Levelling, &c. By Prof. JOHN SCOTT, Editor of the "Farmers' Gazette," late Professor of Agriculture and Rural Economy at the Royal Agricultural College, Cirencester, &c. &c. In One Vol., 1,150 pages, half-bound, with over 600 Illustrations, 12s.

"Written with great care, as well as with knowledge and ability. The author has done his work well; we have found him a very trustworthy guide wherever we have tested his statements. The volume will be of great value to agricultural students."—*Mark Lane Express.*

"For a young agriculturist we know of no handy volume likely to be more usefully studied."—*Bell's Weekly Messenger.*

English Agriculture.

THE FIELDS OF GREAT BRITAIN: A Text-Book of Agriculture, adapted to the Syllabus of the Science and Art Department. For Elementary and Advanced Students. By HUGH CLEMENTS (Board of Trade). Second Ed., Revised, with Additions. 18mo, 2s. 6d. cl.

"A most comprehensive volume, giving a mass of information."—*Agricultural Economist.*

"It is a long time since we have seen a book which has pleased us more, or which contains such a vast and useful fund of knowledge."—*Educational Times.*

Tables for Farmers, etc.

TABLES, MEMORANDA, AND CALCULATED RESULTS *for Farmers, Graziers, Agricultural Students, Surveyors, Land Agents Auctioneers, etc.* With a New System of Farm Book-keeping. Selected and Arranged by SIDNEY FRANCIS. Second Edition, Revised. 272 pp., waistcoat-pocket size, 1s. 6d. limp leather.

"Weighing less than 1 oz., and occupying no more space than a match box, it contains a mass of facts and calculations which has never before, in such handy form, been obtainable. Every operation on the farm is dealt with. The work may be taken as thoroughly accurate, the whole of the tables having been revised by Dr. Fream. We cordially recommend it."—*Bell's Weekly Messenger.*

"A marvellous little book. . . . The agriculturist who possesses himself of it will not be disappointed with his investment."—*The Farm.*

AGRICULTURE, FARMING, GARDENING, etc. 37

Farm and Estate Book-keeping.
BOOK-KEEPING FOR FARMERS & ESTATE OWNERS. A Practical Treatise, presenting, in Three Plans, a System adapted for all Classes of Farms. By JOHNSON M. WOODMAN, Chartered Accountant. Second Edition, Revised. Cr. 8vo, 3s. 6d. cl. bds.; or 2s. 6d. cl. limp.
"The volume is a capital study of a most important subject."—*Agricultural Gazette.*
"Will be found of great assistance by those who intend to commence a system of book-keeping, the author's examples being clear and explicit, and his explanations, while full and accurate, being to a large extent free from technicalities."—*Live Stock Journal.*

Farm Account Book.
WOODMAN'S YEARLY FARM ACCOUNT BOOK. Giving a Weekly Labour Account and Diary, and showing the Income and Expenditure under each Department of Crops, Live Stock, Dairy, &c. &c. With Valuation, Profit and Loss Account, and Balance Sheet at the end of the Year, and an Appendix of Forms. Ruled and Headed for Entering a Complete Record of the Farming Operations. By JOHNSON M. WOODMAN, Chartered Accountant. Folio, 7s. 6d. half-bound. [*culture.*
"Contains every requisite form for keeping farm accounts readily and accurately."—*Agri-*

Early Fruits, Flowers and Vegetables.
THE FORCING GARDEN; or, How to Grow Early Fruits, Flowers, and Vegetables. With Plans and Estimates for Building Glasshouses, Pits and Frames. By SAMUEL WOOD. Crown 8vo, 3s. 6d. cloth.
"A good book, and fairly fills a place that was in some degree vacant. The book is written with great care, and contains a great deal of valuable teaching."—*Gardeners' Magazine.*
"Mr. Wood's book is an original and exhaustive answer to the question 'How to Grow Early Fruits, Flowers and Vegetables?'"—*Land and Water.*

Good Gardening.
A PLAIN GUIDE TO GOOD GARDENING; or, How to Grow Vegetables, Fruits, and Flowers. With Practical Notes on Soils, Manures, Seeds, Planting, Laying-out of Gardens and Grounds, &c. By S. WOOD. Fourth Edition, with numerous Illustrations. Crown 8vo, 3s. 6d. cloth.
"A very good book, and one to be highly recommended as a practical guide. The practical directions are excellent."—*Athenæum.*
"May be recommended to young gardeners, cottagers, and specially to amateurs, for the plain, simple, and trustworthy information it gives on common matters too often neglected."—*Gardeners' Chronicle.*

Gainful Gardening.
MULTUM-IN-PARVO GARDENING; or, How to make One Acre of Land produce £620 a-year by the Cultivation of Fruits and Vegetables; also, How to Grow Flowers in Three Glass Houses, so as to realise £176 per annum clear Profit. By S. WOOD. Fifth Edition. Crown 8vo, 1s. sewed.
"We are bound to recommend it as not only suited to the case of the amateur and gentleman's gardener, but to the market grower."—*Gardeners' Magazine.*

Gardening for Ladies.
THE LADIES' MULTUM-IN-PARVO FLOWER GARDEN, and Amateurs' Complete Guide. By S. WOOD. With Illusts. Cr. 8vo, 3s. 6d. cl.
"This volume contains a good deal of sound, common sense instruction."—*Florist.*
"Full of shrewd hints and useful instructions, based on a lifetime of experience."—*Scotsman.*

Receipts for Gardeners.
GARDEN RECEIPTS. By C. W. QUIN. 12mo, 1s. 6d. cloth.
"A useful and handy book, containing a good deal of valuable information."—*Athenæum.*

Market Gardening.
MARKET AND KITCHEN GARDENING. By Contributors to "The Garden." Compiled by C. W. SHAW, late Editor of "Gardening Illustrated." 12mo, 3s. 6d. cloth boards.
"The most valuable compendium of kitchen and market-garden work published."—*Farmer.*

Cottage Gardening.
COTTAGE GARDENING; or, Flowers, Fruits, and Vegetables for Small Gardens. By E. HOBDAY. 12mo, 1s. 6d. cloth limp.

Potato Culture.
POTATOES: How to Grow and Show Them. A Practical Guide to the Cultivation and General Treatment of the Potato. By JAMES PINK. Second Edition. Crown 8vo, 2s. cloth.

LAND AND ESTATE MANAGEMENT, LAW, etc.

Hudson's Land Valuer's Pocket-Book.
THE LAND VALUER'S BEST ASSISTANT: Being Tables on a very much Improved Plan, for Calculating the Value of Estates. With Tables for reducing Scotch, Irish, and Provincial Customary Acres to Statute Measure, &c. By R. HUDSON, C.E. New Edition. Royal 32mo, leather, elastic band, 4s.

"This new edition includes tables for ascertaining the value of leases for any term of years and for showing how to lay out plots of ground of certain acres in forms, square, round, &c., with valuable rules for ascertaining the probable worth of standing timber to any amount; and is of incalculable value to the country gentleman and professional man."—*Farmers' Journal.*

Ewart's Land Improver's Pocket-Book.
THE LAND IMPROVER'S POCKET-BOOK OF FORMULÆ, TABLES and MEMORANDA *required in any Computation relating to the Permanent Improvement of Landed Property.* By JOHN EWART, Land Surveyor and Agricultural Engineer. Second Edition, Revised. Royal 32mo, oblong, leather, gilt edges, with elastic band, 4s.

"A compendious and handy little volume."—*Spectator.*

Complete Agricultural Surveyor's Pocket-Book.
THE LAND VALUER'S AND LAND IMPROVER'S COMPLETE POCKET-BOOK. Consisting of the above Two Works bound together. Leather, gilt edges, with strap, 7s. 6d.

"Hudson's book is the best ready-reckoner on matters relating to the valuation of land and crops, and its combination with Mr. Ewart's work greatly enhances the value and usefulness of the latter-mentioned. . . . It is most useful as a manual for reference."—*North of England Farmer.*

Auctioneer's Assistant.
THE APPRAISER, AUCTIONEER, BROKER, HOUSE AND ESTATE AGENT AND VALUER'S POCKET ASSISTANT, for the Valuation for Purchase, Sale, or Renewal of Leases, Annuities and Reversions, and of property generally; with Prices for Inventories, &c. By JOHN WHEELER, Valuer, &c. Fifth Edition, re-written and greatly extended by C. NORRIS, Surveyor, Valuer, &c. Royal 32mo, 5s. cloth.

"A neat and concise book of reference, containing an admirable and clearly-arranged list of prices for inventories, and a very practical guide to determine the value of furniture, &c."—*Standard.*

"Contains a large quantity of varied and useful information as to the valuation for purchase, sale, or renewal of leases, annuities and reversions, and of property generally, with prices for inventories, and a guide to determine the value of interior fittings and other effects."—*Builder.*

Auctioneering.
AUCTIONEERS: THEIR DUTIES AND LIABILITIES. A Manual of Instruction and Counsel for the Young Auctioneer. By ROBERT SQUIBBS, Auctioneer. Second Edition, Revised and partly Re-written. Demy 8vo, 12s. 6d. cloth.

"The position and duties of auctioneers treated compendiously and clearly."—*Builder.*

"Every auctioneer ought to possess a copy of this excellent work."—*Ironmonger.*

"Of great value to the profession. . . . We readily welcome this book from the fact that it treats the subject in a manner somewhat new to the profession."—*Estates Gazette.*

Legal Guide for Pawnbrokers.
THE PAWNBROKERS', FACTORS' AND MERCHANTS' GUIDE TO THE LAW OF LOANS AND PLEDGES. With the Statutes and a Digest of Cases on Rights and Liabilities, Civil and Criminal, as to Loans and Pledges of Goods, Debentures, Mercantile and other Securities. By H. C. FOLKARD, Esq., Barrister-at-Law, Author of "The Law of Slander and Libel," &c. With Additions and Corrections. Fcap. 8vo, 3s. 6d. cloth.

"This work contains simply everything that requires to be known concerning the department of the law of which it treats. We can safely commend the book as unique and very nearly perfect."—*Iron.*

"The task undertaken by Mr. Folkard has been very satisfactorily performed. . . Such explanations as are needful have been supplied with great clearness and with due regard to brevity."—*City Press.*

LAND AND ESTATE MANAGEMENT, LAW, etc. 39

Law of Patents.
PATENTS FOR INVENTIONS, AND HOW TO PROCURE THEM. Compiled for the Use of Inventors, Patentees and others. By G. G. M. HARDINGHAM, Assoc.Mem.Inst.C.E., &c. Demy 8vo, cloth, price 2s. 6d.

Metropolitan Rating Appeals.
REPORTS OF APPEALS HEARD BEFORE THE COURT OF GENERAL ASSESSMENT SESSIONS, from the Year 1871 to 1885. By EDWARD RYDE and ARTHUR LYON RYDE. Fourth Edition, brought down to the Present Date, with an Introduction to the Valuation (Metropolis) Act, 1869, and an Appendix by WALTER C. RYDE, of the Inner Temple, Barrister-at-Law. 8vo, 16s. cloth.

"A useful work, occupying a place mid-way between a handbook for a lawyer and a guide to the surveyor. It is compiled by a gentleman eminent in his profession as a land agent, whose speciality, it is acknowledged, lies i the direction of assessing property for rating purposes."—*Land Agents' Record.*

"It is an indispensable wo of reference for all engaged in assessment business."—*Journal of Gas Lighting.*

House Property.
HANDBOOK OF HOUSE PROPERTY. A Popular and Practical Guide to the Purchase, Mortgage, Tenancy, and Compulsory Sale of Houses and Land, including the Law of Dilapidations and Fixtures; with Examples of all kinds of Valuations, Useful Information on Building, and Suggestive Elucidations of Fine Art. By E. L. TARBUCK, Architect and Surveyor. Fourth Edition, Enlarged. 12mo, 5s. cloth.

"The advice is thoroughly practical."—*Law Journal.*
"For all who have dealings with house property, this is an indispensable guide."—*Decoration.*
"Carefully brought up to date, and much improved by the addition of a division on fine art."
"A well-written and thoughtful work."—*Land Agent's Record.*

Inwood's Estate Tables.
TABLES FOR THE PURCHASING OF ESTATES, Freehold, Copyhold, or Leasehold; Annuities, Advowsons, etc., and for the Renewing of Leases held under Cathedral Churches, Colleges, or other Corporate bodies, for Terms of Years certain, and for Lives; also for Valuing Reversionary Estates, Deferred Annuities, Next Presentations, &c.; together with SMART'S Five Tables of Compound Interest, and an Extension of the same to Lower and Intermediate Rates. By W. INWOOD. 23rd Edition, with considerable Additions, and new and valuable Tables of Logarithms for the more Difficult Computations of the Interest of Money, Discount, Annuities, &c., by M. FEDOR THOMAN, of the Société Crédit Mobilier of Paris. Crown 8vo, 8s. cloth.

"Those interested in the purchase and sale of estates, and in the adjustment of compensation cases, as well as in transactions in annuities, life insurances, &c., will find the present edition of eminent service."—*Engineering.*

"'Inwood's Tables' still maintain a most enviable reputation. The new issue has been enriched by large additional contributions by M. Fedor Thoman, whose carefully arranged Tables cannot fail to be of the utmost utility."—*Mining Journal.*

Agricultural and Tenant-Right Valuation.
THE AGRICULTURAL AND TENANT-RIGHT-VALUER'S ASSISTANT. A Practical Handbook on Measuring and Estimating the Contents, Weights and Values of Agricultural Produce and Timber, the Values of Estates and Agricultural Labour, Forms of Tenant-Right-Valuations, Scales of Compensation under the Agricultural Holdings Act, 1883, &c. &c. By TOM BRIGHT, Agricultural Surveyor. Crown 8vo, 3s. 6d. cloth.

"Full of tables and examples in connection with the valuation of tenant-right, estates, labour, contents, and weights of timber, and farm produce of all kinds."—*Agricultural Gazette.*
"An eminently practical handbook, full of practical tables and data of undoubted interest and value to surveyors and auctioneers in preparing valuations of all kinds."—*Farmer.*

Plantations and Underwoods.
POLE PLANTATIONS AND UNDERWOODS: A Practical Handbook on Estimating the Cost of Forming, Renovating, Improving and Grubbing Plantations and Underwoods, their Valuation for Purposes of Transfer, Rental, Sale or Assessment. By TOM BRIGHT, F.S.Sc., Author of "The Agricultural and Tenant-Right-Valuer's Assistant," &c. Crown 8vo, 3s. 6d. cloth. *[Just published.*

"Will be found very useful to those who are actually engaged in managing wood."—*Bell's Weekly Messenger.*
"To valuers, foresters and agents it will be a welcome aid."—*North British Agriculturist.*
"Well calculated to assist the valuer in the discharge of his duties, and of undoubted interest and use both to surveyors and auctioneers in preparing valuations of all kinds."—*Kent Herald.*

A Complete Epitome of the Laws of this Country.

EVERY MAN'S OWN LAWYER: A Handy-Book of the Principles of Law and Equity. By A BARRISTER. Twenty-ninth Edition. Revised and Enlarged. Including the Legislation of 1891, and including careful digests of *The Tithe Act*, 1891; the *Mortmain and Charitable Uses Act*, 1891; the *Charitable Trusts (Recovery) Act*, 1891; the *Forged Transfers Act*, 1891; the *Custody of Children Act*, 1891; the *Slander of Women Act*, 1891; the *Public Health (London) Act*, 1891; the *Stamp Act*, 1891; the *Savings Bank Act*, 1891; the *Elementary Education ("Free Education") Act*, 1891; the *County Councils (Elections) Act*, 1891; and the *Land Registry (Middlesex Deeds) Act*, 1891; while other new Acts have been duly noted. Crown 8vo, 688 pp., price 6s. 8d. (saved at every consultation!), strongly bound in cloth. [*Just published.*

*** THE BOOK WILL BE FOUND TO COMPRISE (AMONGST OTHER MATTER)—

THE RIGHTS AND WRONGS OF INDIVIDUALS—LANDLORD AND TENANT—VENDORS AND PURCHASERS—PARTNERS AND AGENTS—COMPANIES AND ASSOCIATIONS—MASTERS, SERVANTS AND WORKMEN—LEASES AND MORTGAGES—CHURCH AND CLERGY, RITUAL—LIBEL AND SLANDER—CONTRACTS AND AGREEMENTS—BONDS AND BILLS OF SALE—CHEQUES, BILLS AND NOTES—RAILWAY AND SHIPPING LAW—BANKRUPTCY AND INSURANCE—BORROWERS, LENDERS AND SURETIES—CRIMINAL LAW—PARLIAMENTARY ELECTIONS—COUNTY COUNCILS—MUNICIPAL CORPORATIONS—PARISH LAW, CHURCHWARDENS, ETC.—PUBLIC HEALTH AND NUISANCES—FRIENDLY AND BUILDING SOCIETIES—COPYRIGHT AND PATENTS—TRADE MARKS AND DESIGNS—HUSBAND AND WIFE, DIVORCE, ETC.—TRUSTEES AND EXECUTORS—INTESTACY, LAW OF—GUARDIAN AND WARD, INFANTS, ETC.—GAME LAWS AND SPORTING—HORSES, HORSE-DEALING AND DOGS—INNKEEPERS, LICENSING, ETC.—FORMS OF WILLS, AGREEMENTS, ETC. ETC.

NOTE.—*The object of this work is to enable those who consult it to help themselves to the law; and thereby to dispense, as far as possible, with professional assistance and advice. There are many wrongs and grievances which persons submit to from time to time through not knowing how or where to apply for redress; and many persons have as great a dread of a lawyer's office as of a lion's den. With this book at hand it is believed that many a* SIX-AND-EIGHTPENCE *may be saved; many a wrong redressed; many a right reclaimed; many a law suit avoided; and many an evil abated. The work has established itself as the standard legal adviser of all classes, and also made a reputation for itself as a useful book of reference for lawyers residing at a distance from law libraries, who are glad to have at hand a work embodying recent decisions and enactments.*

*** OPINIONS OF THE PRESS.

"It is a complete code of English Law, written in plain language, which all can understand. . . Should be in the hands of every business man, and all who wish to abolish lawyers' bills."—*Weekly Times.*

"A useful and concise epitome of the law, compiled with considerable care."—*Law Magazine.*

"A complete digest of the most useful facts which constitute English law."—*Globe.*

"Admirably done, admirably arranged, and admirably cheap."—*Leeds Mercury.*

"A concise, cheap and complete epitome of the English law So plainly written that he who runs may read, and he who reads may understand."—*Figaro.*

"A dictionary of legal facts well put together. The book is a very useful one."—*Spectator.*

"The latest edition of this popular book ought to be in every business establishment, and on every library table."—*Sheffield Post.*

Private Bill Legislation and Provisional Orders.

HANDBOOK FOR THE USE OF SOLICITORS AND ENGINEERS Engaged in Promoting Private Acts of Parliament and Provisional Orders, for the Authorization of Railways, Tramways, Works for the Supply of Gas and Water, and other undertakings of a like character. By L. LIVINGSTON MACASSEY, of the Middle Temple, Barrister-at-Law, M.Inst.C.E.; Author of "Hints on Water Supply." 8vo, 950 pp., 25s. cloth.

"The volume is a desideratum on a subject which can be only acquired by practical experience, and the order of procedure in Private Bill Legislation and Provisional Orders is followed. The author's suggestions and notes will be found of great value to engineers and others professionally engaged in this class of practice."—*Building News.*

"The author's double experience as an engineer and barrister has eminently qualified him for the task, and enabled him to approach the subject alike from an engineering and legal point of view. The volume will be found a great help both to engineers and lawyers engaged in promoting Private Acts of Parliament and Provisional Orders."—*Local Government Chronicle.*

OGDEN, SMALE AND CO. LIMITED, PRINTERS, GREAT SAFFRON HILL, E.C.

www.ingramcontent.com/pod-product-compliance
Lightning Source LLC
Chambersburg PA
CBHW032149160426
43197CB00008B/831